TOOLS OF WAR

JEREMY BLACK

Quercus

Contents

Introduction

Weapons are the means of war and this book sets out to introduce the key ones in military history. As such, it is also a history of war, but as the entries reveal, weapons take on meaning in their social and political contexts. To the modern mind, the subject of weapons can be very simple. Weapons are invented, they offer an advance in military capability, and military history proceeds as a result of these stages. This is not, in fact, the case. First, for many weapons, it is unclear that invention is an appropriate term, certainly if this is understood as a dramatic change abruptly occurring at a certain moment resulting in clear consequences. The process of invention, instead, is generally more complex and many inventions actually are, in fact, re-inventions. This is true, for example, of working submarines (with inventions in 1776, 1797, 1879), percussion hand grenades (1861 and 1905) and flamethrowers (424 BC, 1910 and 1940s); although it can also be argued that all the re-inventions were new.

Furthermore, there is frequently a contrast between the period of invention, the period in which a weapon worked effectively for the first time, and, on the other hand, the period in which military thinking changed in order to take advantage of what the new device allowed the operator of it to do that he could not previously do. There is also the question of how effectiveness was and is defined – it was not one necessarily solved with the initial invention of the weapon. For example, the great constructional problem with the breech-loading rifle was the escape of gas at the breech, and this was the cause of the major delay in its adoption in the 19th century. Similarly, the tendency of the Gatling gun, an early machine gun, to jam delayed its large-scale introduction.

There is also the issue of production. To create an effective weapon which then leads to a relevant change in tactics and, even, doctrine, is only part of the story. It is also necessary to be able to manufacture large numbers of a new weapon, and at a consistent standard, in order to replace losses both in victory and in defeat, and to provide the resources for new operational opportunities. These were all aspects of the process of invention. With firearms, for example, it was necessary to deploy a number of production technologies. Weapon, ammunition, ignition, and propellant all had to be considered. Furthermore, it is important to be able to supply and repair weapons as an example of their functionality and fitness for purpose. New weapons posed particular problems of supply and repair.

There is also the question of how weaponry was perceived at the time. Within Western states, the institutionalization of military systems, particularly for artillery from the 18th century, ensured that the utility of new weapons and systems was carefully probed. Thus, in 1848–9, British officers commanding regiments with rifle companies reported to the Secretary of the Military Board on the best form of rifle ammunition. Investigation, however, did not necessarily ensure appropriate ammunition, or, more generally, usage. As another example, automatic rifles, around since the end of the 19th century, were, when tested by the British during World War I, viewed as nothing more than weapons in which the burden of opening and closing the bolt had been automated. That is not how automatic rifles are viewed today. These are examples of the complexity of a fascinating and important story, one that is outlined in the following pages.

I am most grateful to Wayne Davies, Kelly Devries, Gervase Phillips, John France, Gary Sheffield, David Trim and Everett Wheeler for comments on sections of an earlier draft. It is a great pleasure to dedicate this book to Mike Mosbacher, a supporter of my work and a thoroughly good fellow.

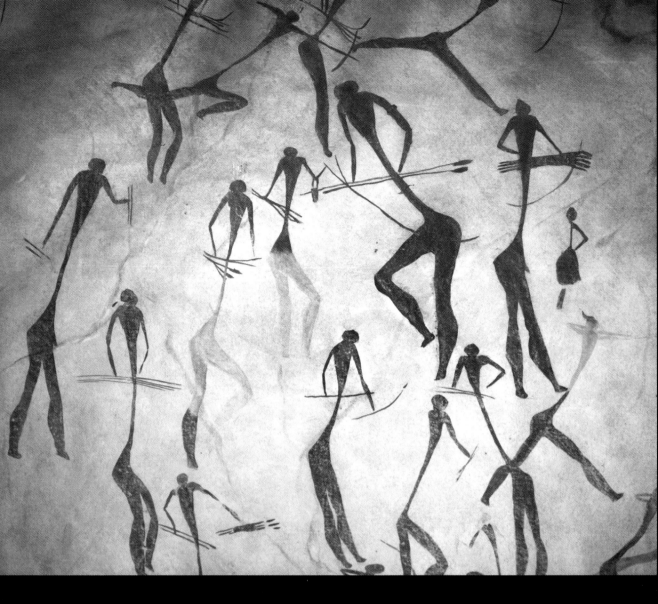

Stone and Metal

'Accurst be he that first invented war.'

CHRISTOPHER MARLOWE

THOUSANDS OF MILLENNIA AGO when our ancestors first roamed the great plains and forests of the world, they were – without weapons other than their hands and teeth – intensely vulnerable to nearly all other animals. Moreover they also needed to kill them in order to get food. Weapons were crucial both to fend off predators, such as bears and wolves, who attacked them, and also to become more successful predators themselves. Early weapons were based on stone that had been worked with other stones in order to make it more effective: chipped in order to create killings points. Spears and arrows were originally stone-tipped. As a result around the globe, hunter-gatherers became more successful and more dominant in the animal world in the prehistoric period.

The dates of developments in weaponry are necessarily imprecise, but they stemmed from the growth of a tool-based culture. In about 100,000 BC, stone tools, shaped by striking flakes from the core, started to be made. Humans were able to make successful weapons, especially composite tools – points and blades mounted in wood or bone hafts – which were developed in areas of early settlement, such as Israel, about 45,000 BC. Bows and arrows, harpoons and spear throwers were used in Europe from about 35,000 BC. Clovis points, made by chipping rocks into sharp, flat shapes, in order to produce large stone points which were able to pierce the hides of mammoths, for example, were used from about 10,000 BC in North America. They have been found there in numerous sites.

HUNTING HERDS

Weapons alone were not the key. Humans also had important physiological and social advantages over other animals. They could perspire and move at the same time, a major benefit in both pursuit and fighting. Other animals, in contrast, stopped to perspire, and were therefore more vulnerable. The human capacity to communicate through language was also very valuable. This was linked to their ability to organize into groups, a vital skill when hunting herds of huge animals such as mastodons and mammoths, although animals such as lions and wild dogs shared this ability.

Humans were also able to develop their tools, which was to become a key characteristic in the history of weaponry and war. They tested the opportunities of everything they could lay their hands on – stone, wood, bone, hide, antler, fire and clay, both on their own and in competition. About 10,000 BC, for example, the Japanese began to use bows and arrows, which gave greater penetrative power than the spears and axes previously thrown at other animals.

The improvements in temperature at the close of the Ice Age further enhanced the situation for humans: the animals they hunted, such as deer, became more plentiful as it became warmer. Some animals also fell victim to humans: the mastodons and mammoths were wiped out – in Europe by about 10,000 BC, and in North America by about 9,000 BC. Furthermore, humans were increasingly able to confront other carnivores, such as bears and wolves, which reduced the competition they posed for food, as well as their threat to humans. The human population grew because of the better climate and their development of agriculture which all helped their potential to succeed. Bears, wolves and other competing carnivores were gradually driven away from areas of human settlement and into mountain and forest fastnesses.

FULL-TIME WARRIORS

The role of agriculture is a reminder that key weapons relate not only to what is used in fighting, but also to the tools that increase society's capabilities. In the case of agriculture, food improved nutrition, but also ensured a surplus that eventually permitted specialization in human activity in the shape of full-time warriors. Humans moved from harvesting wild cereals, which they did in the Middle East from at least 17,000 BC, to cultivating crops. This became large scale in western Asia and north China by about 7,000 BC, in Egypt by 6,000 BC, and in northern India and central Europe by 5,000 BC. The spread of agriculture accentuated the development of permanent settlements and led to important innovations in irrigation and in the processing and storage of food. Metalworking and trade both became important, as food surpluses made it possible for some workers to concentrate on other tasks. In the long term, this was to lead to specialization in military service, as opposed to a situation in which all or most fit males fought.

Conflict involved not only fighting with other animals, but also increasingly with other humans. Archaeological finds consist of weapons, defences and also marks on human skeletons. This evidence is extremely valuable, but does not explain the motives or goals of conflict, and, without a knowledge of these, it is difficult to evaluate military effectiveness and therefore weaponry. For example, skeletal remains showing violence as a cause of death may indicate war, but may also indicate murder or feud.

There has been debate about the propensity of early peoples for conflict, and about the extent to which this was unlimited. Discussion has focused on whether early warfare contained crucial limiting ritual and symbolic elements. There was a strong tendency in the 1960s to eulogize early humans and to blame warfare on the development of divisive hierarchies, but this account of a primitive virtue that became corrupted by society is difficult to believe in.

Fighting, indeed, is not some result of the corruption of humankind by society. Instead, it is integral to human society – the Garden of Eden is only a myth. If, from the outset, humans competed with other animals, and fed and protected themselves as a result of these struggles, there was far less contrast between this and fighting other humans than in modern culture. Indeed, the pattern of doing both, and celebrating both in ritual and culture, in longstanding modern hunter-gatherer societies, such as those in Amazonia and New Guinea, indicates a situation that was formerly far more common.

> Among the indigenous population of North America in pre-'contact' (i.e. before the arrival of European settlers) times, there appears to have been no sharp distinction between raiding other human groups and hunting animals.

For example, among the indigenous population of North America in pre-'contact' (i.e. before the arrival of European settlers) times, there appears to have been no sharp distinction between raiding other human groups and hunting animals. The two activities merged. In part, this may be because non-tribal members were not viewed as human beings, or at least not as full persons. Although the context was very different, the presentation and treatment of enemies as beasts or as subhuman can also be seen in the case of some conflict by both modern and earlier states, and indeed becomes a key way to raise support for such conflict.

To return to the Native Americans, it is very difficult to define what war meant to them. Instead, there was both 'public' warfare, in the form of conflict between tribes, and 'private' warfare, raids with no particular sanction, often designed to prove manhood, as well as hunting. The distinction between public and private war was also a problem elsewhere, for example in early Rome. There, the *gentes* (clans) could raise their own armies from their members and clients, and conduct their own wars.

In 7,000–5,000 BC in both West Asia and southeast Europe, it was discovered that heating could be used to isolate metals from ore-bearing metals. Soft metals, which melt at low temperatures, were the first to be used, and this explains why copper was the basis of metal technology before iron. The Stone Age began to be replaced by the successive ages of metal, but the concept of a revolutionary change in this process is problematic. There was a considerable overlap of flint tools (including weapons) with copper, copper with bronze, and bronze with iron, rather than a sudden and complete supplanting of one technology by another. In addition, the metallurgical aspects of making and processing the metals and alloys were not static and different processes developed in particular parts of the world. In the third millennium BC, bronze, which was made by alloying copper with tin, was widely adopted as it was stronger and more durable than pure copper.

Metals offered greater potency, not least because they provided stronger penetration and weight, the key requirements for success in hand-to-hand conflict, with the additional factor of the reduced bulk necessary for ease of use and mobility. Metal swords probably developed in Europe in the second millennium BC. These swords eventually became both cutting and thrusting weapons.

HAND-TO-HAND FIGHTING

Some of the early weapons continued into recent times, in part because they were effective in hand-to-hand fighting. This was true, for example, of the bone clubs used by the Maori of New Zealand, who only adopted firearms when exposed to European pressure in the early 19th century, and of the wood, bone or stone weapons used by the aboriginal peoples of Australia and Siberia. The former, however, were conquered by Europeans in the late 18th and 19th century, and the latter in the 17th century. The native societies of the Americas relied on wood and stone, not steel and iron, and this gave the Spaniards superiority over their Aztec and Inca opponents in the 16th century in modern Mexico and Peru respectively: metal weapons were more effective than stone. Metal is also more malleable than stone. However, when a small Spanish force fought natives on the Yucatán coast of modern Mexico in 1517, it was driven off by warriors carrying stone-tipped spears, and flint or obsidian-edged swords. The continued use of such weapons into recent times serves as an important reminder of the folly of assuming that because a weapon becomes redundant in one part of the world, it does so everywhere else.

> Metals offered greater potency, not least because they provided stronger penetration and weight, the key requirements for success in hand-to-hand conflict... Metal swords probably developed in Europe in the second millennium BC.

Shields

'The most common and simple defensive
armament carried by soldiers from
prehistory to the end of the Middle Ages
was the shield. Prehistoric cave paintings
portray wooden shields carried in the hands
of hunters as a means to ward off attacks.'

KELLY DEVRIES 'SHIELDS' *THE OXFORD COMPANION TO MILITARY HISTORY*
ED. RICHARD HOLMES.

SHIELDS PLAYED A KEY ROLE IN COMBAT while it remained hand-to-hand. They derived from the need to protect as well as to attack. Such protection could be offered by the same weapon being used for attack and defence, as with swords, which could be used to parry as well as thrust or slash; but, in defence, this was less effective than a weapon that would provide a larger blocking capability.

As men have two arms, it was possible to wield both an attacking and a defending weapon simultaneously, an option that would not have been possible had humans been reliant solely on weapons that required two arms, such as bows and arrows, slings, large axes and pikes. Shields, indeed, could be readily combined by individual soldiers with their use of many hand held weapons, notably swords and javelins. They also provided a measure of protection against missile weapons such as arrows and javelins. Moreover, a spike added to the shield's boss turned the shield itself into an offensive weapon at close quarters.

The evolution of the early shield is unclear, but reflected an adaptation of natural materials, such as wood and hide. Wooden shields were depicted in prehistoric cave paintings. Subsequently, as the range of worked material increased, so shields could be made from more substances, and it was also possible to arrive at more combinations. The use of beaten bronze was important in Classical Greece, although it was no easy task to fit a bronze facing to a wooden core. The bronze could also be decorated to show status.

The shield-bearing armies of Classical Greece were a key example of the extent to which economic development led to state forms that could support more sophisticated and better-armed militaries. Economic development was linked to social composition, especially the emergence of powerful élites that provided political direction. States followed, with, for example, Narmer uniting the towns along the lower Nile in about 3,100 BC. The wish to control, and clashing interests, encouraged the walling of settlements and large-scale conflict. In the north China plain in the third millennium BC, fortified settlements and metal weapons appeared.

The density of population and the nature of economic, social and political organizations, also greatly affected military structure and warfare. Natural and human environments combined to ensure a variety of military systems. In part, this was seen in the lack, in many regions of the world, of state-directed regular forces until comparatively recently. This owed a lot to the absence of any powerful sovereign authority across much of the world. It is more appropriate to think of tribal and feudal groups, rather than a state-centric system with large armies using uniform weaponry.

KEY VARIATIONS

The use of shields by many different societies is striking. The size of shields varied greatly, as did their grip, and both were important for their capability as well as their tactical use. Key issues included the shield's weight, which was a problem both on the march and during battles. If the weight was too great, then the resulting strain on the arm might create a problem not only for holding the shield, but also for the use of weapons by the other arm. Weight was related to material and also to size. The larger the size, the greater the ability to deflect blows or missiles, but there was also a loss in terms of mobility, manoeuvrability, and the capacity to use offensive weapons simultaneously. For cavalry the load-bearing capacity of horses posed a different issue.

Key variations in the grip included the central handgrip seen with many shields. Among the variations were the round shields, that possibly began in Central Europe, in which the shield was closely attached to the left forearm, which was put inside the arm band. In contrast, there was also the positioning of the grip near the rim, with a strap that was held in order to prevent the shield moving down the forearm.

The pre-hoplite Greek shield was a relatively small circular shield with a central single grip, and was suspended from the neck by a thong. The shield could be swung around to cover the back when in retreat. In contrast, the shield of the hoplites – literally meaning armoured men – of Classical Greece was much larger, being about one metre (three feet) in diameter, and had a central armband and a handgrip on the inside near the right edge. These shields, which offered protection from chin to knee, were, however, too heavy to slide along the arm and required the full arm and two grips to lift them.

ROMAN LEGIONS

The Roman republican legionary shield was a long oval with a single grip. The switch to a rectangular shield occurred about the time of the Emperor Augustus in the early first century AD, and lasted about a century. These shields were wooden, with the rim reinforced with bronze. The grip was horizontal and was protected by a bronze or iron boss. In the later Roman empire, however, there was a shift away from the weaponry of the legionaries used in the early empire. In the case of shields, about the time of the Emperor Hadrian (AD 117–38), there was a general switch back to an oval shield, although a few rectangular shields still show up in excavations. Oval shields may have been seen as more effective against 'barbarian' attackers, although it is necessary to use care in explaining changes when there is scant, if any, evidence for why decisions were taken.

There was important continuity in shield use between the later Roman empire and the Middle Ages. Shields were made of strips of wood. These were sometimes covered by leather, and strengthened by a metal rim, and many had a metal boss. Size and specifications varied, but the emphasis from the eighth to the 11th centuries was on round shields, after which they became kite-shaped for both infantry and cavalry, until, in the 13th century, there was a new emphasis on smaller, more manoeuvrable, triangular shields. The type and usage of shields varied considerably between cavalry and infantry, although many horsemen fought dismounted.

The value of shields was always challenged by the plunging possibilities of missile fire, although they could be raised to lessen the impact of this fire. Javelins were a problem, but arrows more so, and this was seen with the success of the Normans over the Saxon shield wall at Hastings in 1066. At the end of the Middle Ages and into the 16th century, the pavise – a shield that protected the whole body – was of great importance in protecting archers and gunners.

Shields were rendered redundant by firearms, despite curiosities such as the pistols that protruded through shields purchased for Henry VIII of England in the 1540s. Shields, nevertheless, remained important for forces that did not have to face firearms. For those who had to do so, there was still continued interest in means for protecting soldiers. For example, Ian Hamilton, the Inspector General of British Overseas Forces, proposed wheeled shields for the infantry in the 1900s.

THE KILLING GROUND

In practice, entrenchments were the defensive option that was taken. They enabled soldiers to aim and fire without having to hold a defensive weapon as well. In attack, armoured vehicles offered some substitute for shields, but there has never been an effective source of protection in crossing what is termed the 'killing ground'.

Shields were largely used in the late 20th century by paramilitary forces confronted by urban opponents who were not employing firearms. Indeed, they proved particularly useful against petrol bombs, which became a weapon of choice for urban rioters from the late 1960s, used, for example, in the Paris riots of 1968.

In Northern Ireland, shields were used as a key aspect of riot control, although the challenge there soon became greater than that simply from demonstrators. The police and military response there was important in the defeat of the attempts by the Irish Republican Army (IRA) to overthrow British authority. Fighting between Catholic civil rights activists and the police force of the government of Northern Ireland began in 1968. In the summer of 1969, there came a breakdown in law and order, with open communal violence. The Labour government of Britain was determined not to use troops to maintain order, but, in the face of rioting in Derry and Belfast, did so. However, their very presence became an issue and a cause of violence. The Provisional IRA, founded in 1970, pushed terrorism up the political agenda. The British government reacted with a determined attempt to re-impose control. In Operation Motorman, the IRA's 'no go' areas in Derry and Belfast were reopened for military and police patrols. This led the Provisional IRA to abandon attempts to stage a revolutionary war and, instead, to turn to terrorism.

> Shields were rendered redundant by firearms, despite curiosities such as the pistols that protruded through shields purchased for Henry VIII of England in the 1540s.

RIOT SHIELDS

Paramilitary policing was introduced to mainland Britain from Northern Ireland in response to serious problems of law and order. It proved necessary to use such policing, which included riot shields, in response to large-scale urban rioting in 1981, and also as a result of the bitterly divisive miners' strike of 1984–5. Violent and intimidatory picketing, by the striking miners, was countered by the use of massive police resources in order to keep coal supplies moving and the crucial Nottinghamshire pits open. The police benefited from the establishment of a national system to allocate police resources.

At the military level, the emphasis in recent decades has not been on individual hand-held shields, as with the police, but rather on force protection through field fortifications. There is, however, a fundamental problem, as also with armour, namely that of compromising mobility and therefore limiting the ability of soldiers to discharge their tasks.

Body Armour

'Now my armour and my lyre – its wars are
over – will hang on this wall.'

HORACE

ORIGINALLY, MEN FOUGHT WITH NO SPECIAL CLOTHES intended to protect them from the teeth, blows and weapons of animals and other men. Need and opportunity, however, led to body armour. The two crucial elements were the helmet, designed to protect the head, and the cuirass – a breastplate and backplate fastened together – intended to shield the torso, although other parts of the body were also protected, in particular the limbs by leg and arm guards.

Like other aspects of war, armour involved trade-offs. In the case of armour, these entailed a range of issues. Greater protection, in terms of what was covered, or the strength of the armour, or both, led to problems in the shape of weight and construction. Each affected the ability to move and to fight. Similar problems of trade-off were to recur with tanks, battleships and other weapons.

There were also issues of cost, which, in part, reflected the material used, although the quality of workmanship was also an issue. Thus, the muscled cuirass made of bronze provided good articulation, but was expensive and therefore only used by senior figures. In early Greece and Rome, soldiers provided their own equipment. More armour signified greater wealth and the social distinction of the warrior, rather than being a military necessity. Reputation played a major role in the armour adopted.

In contrast to bronze armour, linen cuirasses were more common, becoming, for example, the standard armour of the hoplite by the fifth century BC. By the time of the Persian Wars, early that century, hoplites were lightening up on the amount of armour worn. Such cuirasses were inexpensive, light and flexible, although these factors could be altered to provide additional protection in the shape of metal scales or plates. This is a reminder of the extent to which there were no hard-and-fast contrasts in armour.

HELMETS

As a result of the varied factors that affected the use of body armour, it is inappropriate to think in terms of some simple model of improvement. Instead, it is instructive to note how different societies responded to these shared problems, sometimes in a similar fashion and sometimes very differently. With helmets there was the question of how much of the face should be covered. There was also the question of the articulation of the pieces, for example whether the helmet had fixed or hinged

Body armour

Fifth century BC
Greek hoplites wore linen cuirasses

Third century BC
The Romans wore chain mail: *lorica hamata*

First century AD
Lorica segmatata: armour of segmented plates introduced by the Romans

Middle Ages
Knights wore plate full body armour – especially in tournaments

16th century
Spanish conquistadors adopted cotton-quilted armour in South America – copied from the indigenous peoples

17th century
Cavalry soldiers wore breast- and backplates

1761
Third Battle of Panipat near Delhi: invading Afghan cavalry wore body armour

20th century
Metal helmets and lightweight flak jackets made of synthetic materials introduced

cheek pieces. For cuirasses, there was also the case of what was considered appropriate for infantry and what for cavalry. Full-length cuirasses for cavalry were very broad in the hips to allow the soldiers to sit on a horse.

After its introduction in the third century BC, *lorica hamata* (mail) became a popular form of Roman armour, although mail shirts were heavy, weighing 12–15 kg. Mail was worn by Roman legionaries until the first century AD after which, although mail was never totally abandoned, *lorica segmentata*, an armour of segmented plates held together with hinges, straps and buckles, offered greater flexibility. In turn, in part in response to the challenge posed by 'barbarian' opponents, or possibly as a consequence of soldiers' opposition to the weight of their body armour, there was a shift from the third century AD away from heavy armour and towards mail. The use of heavy armour, instead, was focused on those cavalry that relied on shock action.

This remained the case during the Middle Ages, with armour providing knights with a measure of protection against the blows of their opponents, whether they fought on foot or on horseback. Mail shirts were the key form of body armour, but they were supplemented by other defences including helmets, shields and plate body armour. The weight of a full body-set of plate armour ensured that it was generally not appropriate on campaign. However, plate armour that focused on protecting vulnerable parts of the body, such as the elbows, kneecaps, shoulders and legs, was important. Such armour offered only limited protection against arrows and pikes, but it did provide some – obviously arrows were most successful against exposed parts of the body, such as the face, and the use of armour lessened the general effectiveness of arrow fire.

TOURNAMENT KNIGHTS

Body armour became increasingly problematic with the onset of firearms, although this was very gradual and armour could deflect or resist bullets. Armour still remained valuable for ceremonial functions as seen in particular in the stylized conflict associated with the tournament.

The heaviest armour for knights came at the end of the Middle Ages when firearms had already begun to render armour obsolete. Nevertheless, it is an important, and insufficiently appreciated, point for understanding the development of European tactics in the Italian Wars (1494–1559), and the replacement of bows by firearms, that armour actually reached its apogee, in terms of protection, in the early to mid-16th century; although the Swiss pikemen generally did not wear armour in order to maintain their mobility. It was really only after the development of the heavy musket that firearms could penetrate armour at effective distances, and armour was then increasingly abandoned.

The limited armour that was used did, however, have value as far as European transoceanic warfare was concerned. For example, the Spaniards benefited in the 16th century in conflict with the Aztecs and Incas. Metal armour, particularly the steel helmets they used, offered more protection than the cotton-quilted armour of their opponents. On the other hand, the Spaniards found the native quilted cotton armour more appropriate for the climate than metal armour. This quilted cotton armour offered the Spaniards protection against spears and bows, although they retained their metal helmets, which were useful against slingshots.

> The emphasis on firepower led to the inevitable demise of infantry armour in Europe from the late 17th century.

This served as an anticipation of the more general problem for Europeans of deciding how best to fight in often very different environments.

ARMOURED CAVALRY

Although full plate armour eventually ceased to have military value in Europe, armour, nevertheless, remained significant in the 17th century. This was particularly so for cavalry, as armour was still valuable in limiting the effects of sword blows. Indeed, an analysis of wounds suffered by the French army in the 1560s and 1570s suggests that 54 per cent were from sword injuries; a figure that in part reflected the important role of cavalry in the Wars of Religion. Seventeenth-century heavy cavalry, or cuirassiers, wore armour that reached to the knee, or the following century, breastplates, while, in the 17th century, light cavalry tended to wear breast- and backplates.

Cavalry outside Europe also wore armour. At the Third Battle of Panipat, fought near Delhi in 1761, the invading Afghan forces that were victorious over the Marathas consisted largely of heavy cavalry equipped with body armour.

In Europe, pikemen also wore breastplates in the 17th century. This provided a degree of protection in infantry combat, as did helmets. Many Chinese soldiers that century and the next also wore mail coats, as did the Bugis of south Sulawesi in the East Indies who were particularly notable as a martial race.

The emphasis on firepower led to the inevitable demise of infantry armour in Europe from the late 17th century. Function was, however, not the only element in military clothing. Instead, there was also a considerable emphasis on display in the uniforms given to soldiers. They were intended to depict the authority and power of the state, and to intimidate domestic and foreign opposition. This encouraged not only the use of uniform but also its non-utilitarian aspects such as prominent headgear, for example that worn by hussars in the Napoleonic period.

Infantry armour returned, however, in the 20th century. Metal plates were used in World War I, although not very widely. Steel helmets proved more important in this war than body shields and steel breastplates, although both the latter were used.

FLAK JACKETS

More significant was the development after 1945 of what were termed 'flak jackets', particularly by the Americans. These were designed to provide protection, but also to avoid the weight problems of metal. Instead, there was an emphasis on plastics, ceramics and new materials, such as the aramid fibre marketed as Kevlar. The use of such armour reflected the greater emphasis placed on avoiding casualties, which was a particular feature of the American army. The ability of modern surgery and medicine to save badly wounded soldiers was also an additional factor.

The revival of body armour was also a product of the possibilities created by the new materials. For example, permeable protective clothing systems have advanced considerably and have been tested on the battlefield.

Armour is not used by the irregular forces that dominate much conflict in the Third World. Indeed the tactical challenge they pose in part rests on the degree to which they cannot be distinguished from the rest of the population. To achieve anonymity they do not use uniform or body armour, although cost considerations also explain why specialized uniform is not worn. Looking to the future, this diversity is likely to remain the pattern.

Chariots

'Some put their trust in chariots,
and some in horses.'

BOOK OF COMMON PRAYER

CHARIOTS WERE SEEN AS A KEY MILITARY ELEMENT IN THE ANCIENT WORLD. The domestication of animals – notably horses – was the crucial prelude to the use of chariots and, indeed, to a widespread expansion in the operational and tactical flexibility of armies. This was denied to societies, such as those in the Americas and Australasia, that lacked the horse. Elsewhere, the horse was the fundamental technology, opening up a range of possibilities. Long before the coming of stirrups, most of these possibilities had already been explored with success: the Scythians were feared horse-archers, and the Sarmatians had heavy cavalry.

However, there were important environmental constraints in the development of cavalry, particularly with disease and terrain for example, horses could not be used in the extensive tsetse-fly belt of Africa nor in the mountainous terrain of Norway.

The development of wheeled transport was also important. The beginnings of the wheel are unclear, and possibly stemmed from log rollers. Wheeled vehicles were in existence in south-west Asia by 3,500 BC. Bronze Age societies had horse-drawn carts and, from about 1700 BC, lighter chariots requiring only two animals were employed. Chariots were prominent in the Near East in the Middle and Late Bronze Age, while, in Mycenaean Greece and Iron Age Britain (700 BC–AD50), the powerful were buried with their chariot and spear.

Chariots proved effective as part of combined weapons systems. In China, the use of chariots, composite bows, and bronze-tipped spears and halberds, developed in the second millennium BC. This is an important reminder that chariots were not only used in the Classical world based on the Near and Middle East. By the third century BC, however, the rise of mass armies, a product of population growth, and the introduction of conscription, ensured that chariots no longer played an important role in China.

TWO-WHEELED CHARIOTS

The combination of the compound bow with the light, two-wheeled chariot, beginning in the 17th century BC in the Middle East, has been seen by some commentators as a tactical revolution that ushered in mass confrontations of chariots carrying archers in the later Bronze Ages; although, at the same time, it is important to avoid an account of military history in which the nature of the weaponry determined success. The Egyptians learned chariotry from the Palestinian Hyksos, who conquered Egypt at the end of the Middle Kingdom (c.2040–1640 BC). Impressions of chariotry can be gained from Egyptian temple reliefs of the Late Kingdom (c.1550–1070 BC) which show a use of bowmen mounted on chariots, as well as of infantry using swords, battle-axes, and other cutting and hitting weapons.

Warfare between states was particularly intense in the Middle East, where a number of cultures clashed soon after the development of cities. The city of Uruk in Mesopotamia developed from about 3,500 BC. Mesopotamia and Egypt supported a series of states, each of which sought to defeat local rivals, and then to expand. The first empire in western Asia was founded in about 2,300 BC by the legendary Sargon the Great, who united the city states of Sumer (southern Mesopotamia) and conquered neighbouring regions, including Elam (southwest Persia) and southeastern Anatolia. The empire collapsed, in large part due to an extended drought, and the Gutians took advantage of the resulting disorder. An empire based on the city of Ur followed (the Ur III empire), and later the Babylonian empire of Hammurabi.

DARING VICTORY

In the 15th century BC, Egypt benefited from the effective organization and arming of its forces in challenging the Mitanni empire of Mesopotamia for dominance of the region west of Mesopotamia proper, the climax being the dramatic, daring victory of Thutmosis III of Egypt over a Syrian coalition at Megiddo in about 1460 BC. The wings of the Egyptian forces enveloped their opponents, and the combination of charioteers with archers won the day. In the 13th century BC, however, Egypt had to give ground before a revitalized and expanding Hittite kingdom (based in Anatolia) that asserted its dominance in the Syrian region.

No fewer than five huge Egyptian bas-reliefs proclaim the victory of Rameses II of Egypt at the Battle of Kadesh in about 1285 BC. In reality, the Hittites ambushed his army with a sudden attack by chariots, and he was lucky to escape. Rameses narrowly avoided death, despite his propagandist claim of victory, in a battle in which large numbers of chariots were used on both sides. The bas-relief monument at Thebes in Egypt depicts Rameses at Kadesh as a chariot rider, indicating the prestige of the role.

In about 1260, the two powers negotiated a treaty acknowledging Hittite expansion and establishing a zone of influence for the two kingdoms. Further afield, both had been expansionist. The Hittites had destroyed Babylon in a raid in 1596 BC, while Egyptian forces also operated south into Nubia which was conquered on several occasions, including in about 1965 BC and c. 1492–71 BC, with its frontier established at the Fourth Cataract on the Nile in c.1446 BC. Both the Egyptians and the Hittites used bronze weapons. The Hittites' precocious, but very limited, use of iron was not a significant factor at this stage.

The Hittite empire disintegrated at the end of the Bronze Age in about 1200 BC, following attack by the mysterious 'Sea Peoples'. This was an aspect of a more widespread collapse, also seen in the fall of Mycenean Greece, Troy, and the Syrian and Canaanite cities, that appears to have been triggered partly by invaders and rebels, and the resulting crises in international trade and political control.

ASSYRIAN EMPIRE

In turn, the Assyrian empire, founded in 950 BC, used iron weapons and benefited from its great ability to supply horses, on which prominent chariot corps depended. Although they also developed some cavalry, their preference was for heavy chariots, with four rather than two horses, and carrying four rather than two men. This greatly increased the firepower of

Major chariot battles of the Ancient World	1460BC **BATTLE OF MEGIDDO**	1285BC **BATTLE OF KADESH**	605BC **BATTLE OF CARCHEMISH**	547BC **BATTLE OF PTERIA**
	Thutmosis III of Egypt defeats a Syrian coalition	Rameses II of Egypt narrowly defeats the Hittites	The Medes and Babylonians crush the Egyptian army	Cyrus the Great of Persia defeats Croesus, king of Lydia

19

chariot forces, as it was possible to carry more archers and also to continue fighting even after sustaining some casualties.

The Assyrians conquered not only Mesopotamia, where Babylon was destroyed in 689 BC, and Phoenicia – the coast of modern Lebanon – but also Egypt, creating, as a result, the first empire to span from the Persian Gulf to the Nile: in Egypt, Memphis was captured by the Assyrian king Esarhaddon in 671 BC and Thebes by King Ashurbanipal in 663 BC. All wars of the ancient Near East were holy wars – contests of rival national deities as well as human armies, and the Assyrians were no exception. They saw themselves spreading the domain and worship of their god Ashur. Their ferocious style of rule, which involved mass killings, torture and deportation, failed, however, because it bred hatred that fostered rebellions. The debilitating attempt to take and hold Egypt, as well as the rebellion of the Babylonians and the rise of the neighbouring Medes, who aligned with Babylon to destroy the Assyrian capital, Nineveh (near modern Mosul) in 612 BC, were responsible for the downfall of Assyria. The Medes and Babylonians crushed the Egyptian army at Carchemish in 605 BC.

Chariots thereafter continued to play a role within armies, even though it was a secondary role to cavalry. The latter offered greater flexibility, not least in different terrains, were far more appropriate for difficult or mountainous areas, and were less expensive. This secondary role for chariots was particularly relevant for the Medes and Persians who were descendants of Central Asian 'horse people'. The Persians under Cyrus the Great (r. 559–30 BC), who defeated Croesus, king of Lydia at Pteria in 547, and gained control of Egypt and the Babylonian empire (capturing Babylon in 539), used chariots, but cavalry warfare was more significant. The chariots provided a way to disrupt opposing battle lines and, to that end, chariots equipped with scythes on their wheels were particularly successful, although understanding of equestrian factors has helped lead to questions about whether chariots charged *en masse* and were therefore really formidable in battle. Scythed chariots are first on record in the early fifth century BC and were, like elephants, probably more of a 'scare tactic' than an effective tactical option.

VENI, VIDI, VICI

Chariots were used by the Persians against the Macedonians, when Alexander the Great invaded the Persian empire. At Arbela (Gaugamela) in 331 BC, the Macedonians thwarted the Persian chariots and cavalry in part by the use of javelin throwers. This reflected the extent to which mobile attacking forces could be weakened by missile throwers. Well-deployed, well-led and prepared infantry could therefore see off chariot attacks, if necessary by opening up gaps in their formation and channelling the chariots through them.

After this chariots were not central to military culture in Eurasia. The Romans, who did not rely on their use, preferring, instead, to focus on infantry, were able to defeat those who did emphasize chariots, as at Zela (now in modern Turkey) in 47 BC when Caesar crushed Pharnaces, king of Bosphorus, after which he remarked *Veni, Vidi, Vici* (I came, I saw, I conquered). When the Romans under Caesar invaded England in 55 and 54 BC, British chariots proved vulnerable to Roman archers.

Instead, cavalry proved a more formidable challenge to the Romans, as with the Parthian mounted archers who resisted Roman advances to the east, and also their infantry, as in Germany. In 53 BC the Parthians inflicted a major defeat on the Romans at Carrhae where an army of 30,000 led by Crassus were defeated and the legions' colours taken as a final humiliation. By then, the glory days of chariots as weapons were largely a feature of the past.

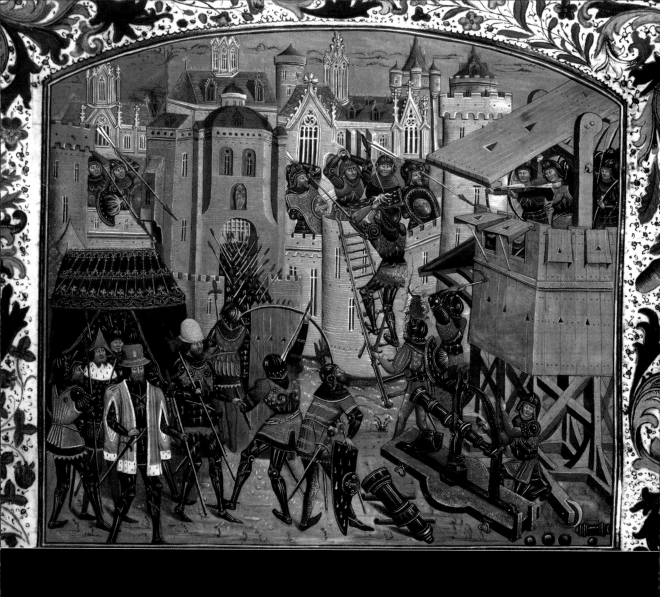

Siege Engines

'The defenders of Breteuil negotiated a
surrender to the King of France, for the siege
engines, which were in action continuously,
were inflicting great damage on them and
they saw no relief from any quarter. They
knew that, if they were taken by assault, they

FORTIFICATIONS OF CITIES INEVITABLY LED TO THE development of ingenious tactics and huge siege weapons which would help them be captured. The wooden horse by which the Greeks allegedly captured Troy is the most famous example. This horse, it has been suggested, was not, however, the vehicle of legend within which the Greek soldiers secretly entered the city, but rather the name given to a covered battering ram. Such rams were certainly illustrated in Assyrian reliefs from the ninth century BC. Battering rams began as tree trunks that were used to exert pressure on a wall or gate. With time, their effectiveness was increased by the addition of an iron head.

These reliefs also indicate that the Assyrians had a sophisticated mastery of siege techniques, including the use of mobile siege towers. Other early siege techniques in the Near East included the construction of an earth mound against walls, in order to enable soldiers to attack without opening themselves to the risk of mounting precarious ladders when exposed to defensive fire. Attempts to undermine walls by mining were also important. So was firing incendiary material into towns in order to cause fires. Starvation of the unfortunate inhabitants was a major alternative if attackers wished to lessen the hazards of combat.

From the fifth century BC in the central Mediterranean, there was the spread or revival of ancient Near Eastern techniques, and attempts to improve on them. This process included the use of large moveable towers designed to provide positions from which the walls could be swept by missiles launched by archers, javelinmen, or machines firing bolts or arrows.

BOLT-SHOOTING CATAPULTS

Dionysius I the Elder of Syracuse (r. 405–367) was particularly associated with the encouragement and use of these revived or new weapons, including bolt-shooting catapults. Made of wood, sinew and horn, these could be equipped with stands and winches, which served to build up a major tensile strength enabling the catapults to outfire handbows. These catapults were used when Dionysius captured Motya from Carthage in 397 BC, and when Alexander the Great successfully besieged Tyre in 332 BC during his conquest of the Persian empire. The catapults were able to provide covering fire for battering rams employed to breach the walls, and also for boarding bridges from which troops moved into the breaches, either from siege towers or, as at Tyre, from ships.

Catapults came in different sizes and threw projectiles that had varied purposes. Large catapults could throw heavy stones that were designed to inflict damage to the structure, for example to the battlements. Medium-sized catapults fired bolts, and lighter ones fired arrows and small stones designed to clear away defenders from their positions. Such anti-personnel weaponry provided an opportunity for gaining tactical dominance and for the use of siege engines against the walls.

By the late fourth century BC, the Hellenistic powers were able to produce more formidable siege weapons, at least judging from surviving accounts. In response to important developments in the scale of fortifications, the siege towers became larger and heavier, able to project more power, and also better defended, for example with iron plates and goatskins to resist the fire missiles and catapult stones launched from the positions they were attacking. The towers were assembleable, so that they could be taken on operations, but others would be made on site if timber was available. The effectiveness of battering rams was enhanced by sheathing them with iron and mounting them on rollers, thus increasing their momentum and accuracy. At the siege of Rhodes in 305–4 BC, there were also iron-tipped borers (made effective by a windlass, pulleys and rollers), that were designed to break holes in the walls. This was a variant on the digging equipment designed to help sappers operate below the walls.

BRIBERY AND TREACHERY

A focus on such ingenious equipment should not lead to a neglect of the more mundane weaponry that was far more commonly used in assaults on fortified positions, such as ladders, nor of the extent to which bribery and negotiation helped lead to the fall of many fortified positions. Treachery was the most effective tool of siegecraft in the Greek world. Also, cities could simply be surrounded and blockaded, and starved into negotiation or surrender, without building siege works at all.

The Romans used similar methods to Hellenistic armies, with siege artillery throwing catapults, bolts and arrows, battering rams employed against walls, ladders deployed to help troops scale the walls, and siege towers and siege mounds used to counteract the height advantage brought by walls. A large ramp of the last type proved crucial in the capture of Masada on the shores of the Dead Sea in AD 74 , which ended the Jewish Revolt that had begun in AD 66. Almost a thousand defenders committed suicide rather than be slaughtered. In AD 70 Jerusalem itself had been successfully taken by the Romans under the Emperor Vespasian's son Titus. The city was to fall again when the revolt by Bar Cochba in AD 132–5 was crushed.

Earlier, the successful siege of Alesia in 52 BC was the last major episode in the Gallic Wars in which Julius Caesar conquered modern France. Alesia was a hilltop stronghold, and Caesar focused on blockade. The siege became a matter of trying to break through the blockade, with the defenders and their relieving army both ultimately unsuccessful. Failure led the garrison to surrender, signalling a final defeat for the Gallic rebels led by Vercingetorix.

Aside from breaching walls by siege artillery, and building towers or mounds, undermining walls was also vital. Roman methods were continued by the Byzantine (Eastern Roman) empire, and effective siegecraft was displayed by its forces at the capture of Palermo in AD 535 and Ravenna in 539, as the forces of the Emperor Justinian (527–65) reconquered much of Italy from the 'barbarian' invaders.

Siege engines were a product not only of technological sophistication, but also of the resources and organization that were so often the key to the development of weaponry. The weaponry used reflected the extent to which some early military powers could wield strong forces.

THE POWERFUL TREBUCHET

The most powerful siege engine in terms of the projectile that could be fired was the trebuchet, which was invented in China in the fifth to third century BC. This was a traction piece that contained a rotating beam placed on a fulcrum. A sling was hung from the beam, and the projectile was placed in it. On the shorter side of the beam, ropes were hung and, when they were pulled, the projectile was thrown forward. This weapon was used by both sides during the Crusades. It was made more effective in the 12th

> Trebuchets hurled stone balls, some of which weighed up to a massive 140 kg (300 lbs) and could cause great damage. They were less inaccurate than might be imagined.

century, when the Arabs replaced the ropes, which had been pulled down by men, with a counterweight. By the 13th century, this had also replaced the earlier model in Europe. Trebuchets were not the only siege engines using a counterweight. The simple Perrier also did the same.

In Europe, trebuchets replaced torsion catapults, as the latter, although more accurate, could not throw such a powerful projectile. Trebuchets hurled stone balls, some of which weighed up to a massive 140 kg (300 lbs) and could cause great damage. Trebuchets were also less inaccurate than might be imagined.

In AD 1266, the impact of trebuchets was amply demonstrated during the lengthy siege of the well-fortified castle of Kenilworth in England. The besieging royal forces brought trebuchets with them, but these were more than counter-balanced by the more powerful trebuchets mounted in the castle by the defenders. As a result, King Henry III had to send for larger machines from London. These, however, did not suffice, and, as an assault also failed, it was necessary to allow disease and starvation to do their work in order to ensure a surrender. Subsequent excavations have shown that some of the balls were hurled 320 metres (350 yards).

In conquering China in the thirteenth century, the Mongols used non-Mongols for siege engineers and for much of the infantry, and Chinese catapult experts proved helpful to them when attacking cities. Aside from bombardment, the Mongols also used blockade in seizing the major cities. The lengthy nature of the sieges, especially of Kaifeng in 1232–3 and of Xiangyang, key to the control of the River Yangzi, in 1268–73, was possible only because of Mongol organization and persistence. When the Xixia empire to the northwest of China was attacked by the Mongols in 1209–10, its capital Ningxia, was flooded when Chinggis (Genghis) Khan dammed the Yellow River.

IVAN THE TERRIBLE

Siege engines remained in use in Europe until the 16th century, being supplemented by and then supplementing cannon, as with Henry V of England's successful siege of Harfleur in Normandy in 1415. However, trebuchets were replaced by gunpowder artillery which offered a mobility and accuracy the trebuchets had lacked. In 1552, in the siege of Kazan, the capital of an Islamic khanate on the River Volga, the Russians under Ivan IV, the Terrible, were victorious, and employed a wooden siege tower carrying cannon and moved on rollers. This was an example of the integration of old and new methods.

By then, nevertheless, the emphasis on artillery ensured that very different tactics were the norm in sieges. Cannon located on the ground were far less vulnerable from counter-battery fire from cannon in the besieged fortress, than cannon in siege towers.

The Turks continued to use siege towers in the 16th century, but at Malta in 1565 they were unsuccessful. A large wooden siege tower was employed against the fortress of Birgu, but it was destroyed by cannon firing chain shot. By the 17th century, the Turks also focused on cannon and the age of siege engines, despite their earlier key role in the fall of fortified positions, was over.

Macedonian Pikes

'Alexander the Great... when he had conquered
what was called the Eastern world... wept for
want of more worlds to conquer.'

ISAAC WATTS

ALEXANDER THE GREAT WAS THE MOST DRAMATICALLY effective general of the ancient world and pikes were crucial to the Macedonian troops he commanded.

The long pike (*sarissa*) was the distinctive weapon of the Macedonians. Their army was developed by King Philip (359–36 BC), who turned Macedonia from a vulnerable region to the north of Greece into the dominant power in the region. This development involved conflict on all the frontiers of Macedonia. Philip could not turn against the Greek cities until he had resolved problems on his northern and western borders, he then moved against the Greek positions, first the cities on the nearby coast, and then lands to the south. This provided conquered territory for redistribution to nobles and helped consolidate Philip's position in Macedonia.

Philip developed the Macedonian army which centred on the phalanx, an infantry column armed with a pike that was sufficiently long to enable several ranks to bring their pikes to bear simultaneously. The pikemen wore enough body armour to provide protection, but not so much as to make it impossible to advance rapidly. This mobility was crucial to the effectiveness of the phalanx. The operation of a unit of pikemen required concerted actions that rested on drill, and therefore on methodical training. If a pike phalanx lost cohesion and became disordered, then it became far weaker.

BATTLE OF CHAEREONEA

This infantry force was matched by a cavalry similarly designed for offensive operations. Frequently in Macedonian tactics, the phalanx played the secondary role. It fixed the opponent, so that the cavalry could deliver the 'knock out' blow. Furthermore, there was a good siege train.

Having negotiated the Peace of Philocrates with the Greeks in 346 BC, Philip then began preparing for war with Persia, the dominant power in Anatolia (Asian Turkey). This led him to intervene in eastern Thrace, which caused anxiety in Athens which obtained its grain from the Black Sea via the Bosphorus and the Dardanelles. In 340, Philip besieged Byzantium and Perinthus, although this resulted in his blocking the Athenian grain trade.

War with Athens was the result and it led, in 338, to the Battle of Chaereonea, in which the forces of Athens and Thebes sought to block Philip's southward advance. It is not clear how precisely the Macedonians won the battle, a characteristic shared by many battles, but the victory led to the conquest of Greece, with most cities now rallying to the Macedonians in the League of Corinth. This was designed to unite Greece behind a Macedonian invasion of the Persian empire. Philip's advance guard landed in Anatolia in 337 BC, but he was murdered in 336 before he could take command there.

His role was taken over by his son, Alexander the Great (r. 336–23 BC), who conquered the Persian empire, an astonishing achievement. Alexander's army invaded the empire in 334 with 15,000 Macedonian infantry as the core unit. The following year, Alexander defeated Darius III at the Battle of Issus. This battle opened the way for Alexander to advance into Phoenicia, Palestine and Egypt. The advance into Phoenicia in 332 and, in particular the successful six-month siege of Tyre, was important to achieve a shift in the naval balance against the Persians, which helped ensure Alexander's dominance in the eastern Mediterranean. Gaza was also captured in 332 after a difficult siege. In Egypt, Alexander was treated as pharaoh and made a journey in 331 to the shrine at the desert oasis of Siwah.

Alexander then returned to Syria and turned east to defeat Darius at Gaugamela (1 October 331 BC) near Nineveh, the decisive defeat of the Persian army. Alexander's force was 7,000 strong, that of Darius 40,000, although much of the latter were weak and poorly trained infantry who lacked the spirit of the battle-hardened Macedonians. Darius relied on his chariots and cavalry, but the chariots were hit by defensive fire from archers and spearmen. The cavalry placed serious pressure on the Macedonians, but Alexander's cavalry itself hit the Persian left, and, with Darius apparently killed (in fact the chariot driver behind him was the victim), many of the Persians fled, destroying the cohesion of their centre. Alexander was primarily a cavalry general.

Alexander then marched east, initially in pursuit of Darius, who was, however, murdered in 330 BC by his general Bessus, who tried to establish himself as an independent ruler, only to fall victim to Alexander. Alexander campaigned in what is now Iran, Afghanistan and Pakistan, reaching Herat in 330 BC and crossing the Hindu Kush in 329, fighting in very different terrains and against unfamiliar opponents, including the elephants deployed against him in India in 326 by King Porus. This advance took Alexander beyond the bounds of Persian power. Alexander's troops, however, refused to go beyond the Hyphasis River.

DEATH OF ALEXANDER

Thwarted of his ambition of advancing down the River Ganges to the Bay of Bengal, Alexander instead advanced along the River Indus, meeting tough resistance, before marching back to Persia across the Gedrosian Desert in 325 BC. Alexander died in Babylon in 323 aged 32. It is unclear what he would have pressed on to do, possibly campaigning in Arabia and against Carthage.

Alexander left no heir and his generals sought to carve realms for themselves. This led to wideranging conflict as they fought among themselves, with their armies in part representing a continuation of Macedonian methods, and in part adopting those of the troops they raised locally. Alexander's army indeed had come to include local troops, first as auxiliaries but eventually in the infantry.

In 301 BC, Antigonus, one of the leading generals, was defeated and killed by Seleucus and Lysimachus at the Battle at Ipsos in Anatolia, only for Lysimachus to be defeated by Seleucus at Corupedium in 281 BC, also in Anatolia. The conflicts of the Hellenistic period after Alexander, produced a *de facto* balance of power between the successor kingdoms of Macedon (ruled by the Antigonids), Syria, including Iraq and Persia (Seleucids), and Egypt (Ptolemies), that was only disrupted by the incursion of Rome in 200 BC immediately

Alexander the Great

336 BC
Alexander proclaimed king at the age of twenty, on the death of his father Philip
333 BC
Alexander defeats Darius III at the Battle of Issus
332 BC
Advances into Phoenicia, Palestine and Egypt where he founds the city of Alexandria in 331
331 BC
Decisive defeat of the Persians at the Battle of Gaugamela
330 BC
Alexander destroys the royal Persian palace of Persepolis
329 BC
The army crosses the Hindu Kush mountain range
327 BC
Invades India
323 BC
Alexander dies in Babylon aged 32. His first wife, Roxane, bears his son shortly after his death – Alexander IV of Macedon

following its defeat of Hannibal, the leading general of Carthage. Already before that, however, much of the far-flung Seleucid empire had fallen to independent Hellenistic kingdoms, particularly, to the east, Graeco-Bactria, which became independent in 250 BC.

ELEPHANTS AND HORSES

Hellenistic warfare was characterized by the development of tendencies already seen in Alexander's campaigning, including the much greater use of siege warfare. Hellenistic armies relied on the different resources of the particular kingdoms, although their cores fought in a similar style. There was a use of elephants, who in practice were more impressive than effective, although in the hands of an experienced general such as Porus at the Hydaspes, elephants could be much more than just a scare tactic. Cavalry, instead, proved more valuable, as at Ipsos. This was the largest of the battles fought by Alexander's successors and each side allegedly had over 70,000 troops. Seleucus made good use of elephants, which frightened the horses, and also probably of horse archers. Antigonus's force ebbed away, finally being shattered when its commander fell victim to a javelin.

The Hellenistic armies focused on the phalanx and their pikes became longer, to reach 5.5 metres (18 feet) long. This was a formidable threat if opponents fought as Hellenistic generals wished, but the phalanx lacked flexibility and proved vulnerable to more mobile rivals, particularly in broken terrain. The problems of advancing across such terrain disordered the Spartan phalanx at Sellasia (222 BC), leading to a victory by the Macedonians under Antigonus Doson. The Romans exploited this vulnerability. Furthermore, like pike-based infantry units in European warfare in the 16th and 17th centuries, the phalanx was vulnerable to the consequences of cavalry conflict on their flanks.

IMPERIAL ARMIES

The size of some of the armies of the period could be formidable. Thanks to imperial power, the relatively low productivity of pre-19th-century agrarian economies was not incompatible with large forces. Surprisingly the constraints that primitive command and control technology and practices placed on centralization did not prevent effective organization over huge territories, as in the Seleucid or Roman empires. In China, in the Warring States period (403–221 BC), improved weapons and the use of mass infantry formations led to some of the largest military engagements yet recorded, although the reliability of the literary sources that record enormous armies and high casualties is in question. The army of Song China was maybe 1.25 million men strong by AD 1041.

These forces were based on sophisticated and wideranging systems for raising and supporting troops. In the Achaemenid Persian empire (c.550–331 BC), for which Alexander was the nemesis, land was granted in return for military service, and was graded as horse-land, bow-land and chariot-land according to what had to be provided. The information was recorded in a census maintained by army scribes. When personal service was not required, a tax had to be paid in silver, which gave the government the ability to move resources more easily. These resources could be employed to pay mercenaries and helped ensure that large armies could be fielded by the Persians, although the figure of about 200,000 soldiers for the Persians in the Plataea campaign in Greece in 479 BC, is generally believed to be far too high and would have produced a logistical nightmare. The latter was far less an issue for the 125,000 troops deployed by the USA for the Iraq campaign of 2003.

Roman Swords

'Without a sign his sword the
brave man draws and asks no omen
but his country's cause.'

HOMER, *THE ILIAD*

SWORDS WERE ONE OF THE KEY WEAPONS OF CONFLICT, for both infantry and cavalry for much of history. Metal swords developed in the second millennium BC and iron swords were used with particular effectiveness by the Assyrians. Swords that could both thrust and slash were especially appreciated and were the key weapon of the Romans. Their infantry used a short stabbing iron sword, the *gladius* (originally a Spanish sword), as well as heavy javelins and shields. The swords were used in a disciplined fashion and their legionaries fought together shoulder-to-shoulder. These weapons and methods proved more adaptable than the pikemen of Hellenistic forces, prefiguring the flexibility of Spanish swordsmen when fighting Swiss pikemen in the Italian Wars (AD 1494–1559).

Indeed, Rome became the leading power in the Eastern Mediterranean, thanks to the ability of its armies to defeat the Macedonian phalanx of spearmen at Cynoscephalae (197 BC) and, especially, Pydna (168 BC). However, these battles were in practice close-run episodes, and it is unclear that the Roman legion was as superior to the phalanx as was claimed. Polybius's claim to this effect was largely Roman propaganda aimed at discouraging further Greek revolts. The Greeks argued that the Romans had been lucky, and this should not be discounted.

Roman cavalry were equipped with a longer sword, the *spatha*, a slashing weapon with a straight blade of 60–70 cm (24–30 inches) that was derived from the Celts or Germans. It is first heard about in the Augustan age as the sword of the *auxilia*.

MILITARY IMAGINATION

It is possible to recapture in some detail the organization, tactics and weaponry of the Roman army, which is encouraging, not least because the military history of so much of the ancient world is obscure in comparison, including that of settled societies, such as Phoenicia. Greece, Macedonia and, particularly Rome helped frame the subsequent military imagination of Europeans. Latin texts, such as those of Vegetius, were seen as providing valuable guides, and practices allegedly modelled on those of Rome enjoyed considerable prestige. This can be shown, for example, in 16th-century interest in Classical models of conflict, in Marshal Saxe's idea in the 1730s and 1740s of reviving the legion, as well as in Napoleon's search for validation by using Roman symbols.

The fighting effectiveness of the Romans was particularly evident in the range of their operations, although this entailed far more than tactical skill and weaponry. The Romans also had impressive manpower, resources, willpower and organization. They were especially notable not only for the training and discipline that enabled them to march at a formidable rate, to deploy in a variety of planned formations, to perform complex manoeuvres on the battlefield, and to perform effectively there; but also for the range of physical and military environments that they mastered. In that, as in roadbuilding, however, they had been anticipated to a certain extent by the Persians.

Having unified Italy, in campaigning that also entailed resistance to Celtic attack from the north, and to invasion in 280 BC by Pyrrhus, king of Epirus, whose army included elephants, the Romans eventually triumphed in the three Punic Wars. These conflicts with Carthage involved fighting in Italy, Sicily, Spain and North Africa, where, in 202 BC in the Second Punic War, Scipio Africanus won a decisive victory over Hannibal at Zama. In the

First Punic War (264–41 BC), the Romans benefited from superior resources, and in the Second (218–01 BC) from the combination of greater manpower and the strategic mistakes made by Hannibal, the Carthaginian general. He was better at winning battles, particularly his dramatic victories at Lake Trasimene (217 BC), and Cannae (216 BC), than at winning the war.

The Romans had a government board called *Duoviri Navales* long before the First Punic War, but warfare with Carthage obliged the Romans to develop their fleet in order to contest the waters off Sicily. The adaptability shown in building up the navy, with the Romans literally copying the design of Carthaginian ships, was also apparent in other aspects of Roman military development, including incorporating the infantry organization of the Samnites, and borrowing javelin and sword types used by their enemies. Carthaginian cavalry exposed the inferiority of Roman and Italian cavalry, and the Roman response was to start using similar cavalry supplied by the Numidians of North Africa.

> In Britain, the disciplined Roman infantry, with its body armour, javelins and short swords, had a major advantage over opponents who had little body armour and lacked effective missile power.

RESERVES OF MANPOWER

The large size of the army of republican Rome was a key resource. It derived from Rome's organization of the peoples of Italy into various citizen and allied statuses, all of which were required to serve in the Roman army. Like the Han Chinese, the Romans believed in a mass army based on the adult males of the farming population. This provided huge reserves of manpower for use against Carthage, and also enabled Rome to fight wars on many fronts. This was crucial given the extent of its empire and the resulting range of its commitments.

In the first century BC, the Romans advanced in many direction: Mithridates, king of Pontus, the leading ruler in Anatolia, was defeated, while Julius Caesar conquered Gaul (France), and the Ptolemaic empire of Egypt was annexed. In the following century, the first AD, Mauretania (in modern Morocco), most of Britain, and much of the northern Balkans, were overcome, although the defeat of the Emperor Augustus's army under Varus in the Teutonburger Wald in Germany in AD 9, with the loss of three legions was a severe blow that encouraged subsequent caution on that frontier. In Britain, the disciplined Roman infantry, with its body armour, javelins and short swords, had a major advantage over opponents who had little body armour and lacked effective missile power.

Dacia (modern Romania), followed in AD 101–6, although this proved a difficult conquest. Successive campaigns showed both the range of Roman power and its offensive capability. By the beginning of the second century AD, the legions contained about 160,000 men, although there were also about 220,000 men in auxiliary regiments, as well as naval forces and tribal semi-irregulars.

The Roman army evolved during the period. By the Late Empire, the longer sword was predominant as part of a re-evaluation of fighting methods to cope with the 'barbarian' armies, a re-evaluation that also affected fighting formations. The extent to which the Romans maintained their traditional formation of one main battle line, with the infantry in the centre flanked by cavalry, and a reserve in the rear, is unclear. Late Roman infantry was probably deployed as a phalanx. The sophisticated infantry tactics of the Republic (described in Polybius, Livy and Caesar, although still imperfectly understood) had long been abandoned. Round shields replaced rectangular in the third century AD, while the first known unit of heavily armed cavalry in the Roman army dates from the early second century AD.

Much of the success of the army also rested on the ability to co-opt assistance from neighbouring peoples. These brought valuable skills with horse and/or bow which the Romans lacked. The Romans benefited in particular from Syrian archers, and Gallic, German (especially Batavian) and Moorish (Mauri) cavalry. The Batavians and Mauri were especially prized. Much of the technical terminology of Roman cavalry manoeuvres was Celtic.

A CRUCIAL CAVALRY WEAPON

Swords were the crucial cavalry weapon in medieval Europe, as well as being important for the infantry. As knights frequently dismounted to fight, they needed to use swords that were appropriate for conflict on land. In the subsequent age of gunpowder, swords continued to be important as infantry weapons and even more for cavalry.

This remained more true of infantry outside Europe. In China, the large force that Zheng Chenggong (known to Europeans as Coxinga) led to the siege of Nanjing in 1659 against the Manchu conquerors was mostly armed with swords: two-handed long, heavy swords, or short swords carried with shields. Swords and knives also were important in the fighting in Tibet involving the Dsungar invaders in 1717. The Gurkhas, who greatly expanded their power along the Himalayan chain in the 18th century, used swords (kukris). By then, while swords continued to play a major role in European cavalry combat, they were only used in infantry conflict by officers.

In the late 19th century, swords fell into disuse as fighting (as opposed to ceremonial) weaponry. This reflected both the adoption of Western military forms by non-Western states, and the declining importance of cavalry in the face of firepower. In 1877, the traditional Japanese military system was defeated by the modern one with the failure of a samurai uprising in the domain of Satsuma. The large samurai force, armed with swords and matchlock muskets, was defeated by the new mass army of conscripted peasants armed with rifles: individual military prowess and bravery succumbing to the organized, disciplined force of an army that, on an individual basis, was less proficient. This episode served as the basis for the 2003 film *The Last Samurai*. Earlier, however, swords had reigned supreme as one of the key weapons in history.

Galleys

'A Galley is a long flat one-decked vessel,
though it hath two masts. Yet they generally
make use of oars, because they are built so as
not to be able to endure a rough sea: and
therefore their sails for the most part are
useless... there are five slaves to every oar; one
of them, a Turk, who being generally stronger
than the Christians, is set at the upper end, to
work it with more strength.'

JOHN BION, ON THE CONDITION OF LIFE ABOARD THE FRENCH GALLEYS, 1703–4

OARED WARSHIPS WERE FOR MANY CENTURIES BEFORE CHRIST WAS BORN, THE STAPLE OF NAVAL WARFARE in the Mediterranean until the 17th century when Northern European warships reliant on windpower came to play a key role. Modern full-scale reconstruction of triremes, a type of galley used for Classical warfare, has helped greatly in understanding the options faced by contemporary commanders. The oarsmen were seated on up to three levels, and between two and 16 men pulled on the oars in each vertical group of oarsmen. Galleys had rams on their prows, which could be used to devastating effect, but the preferred tactic was bombard and board, using catapults, archers and javelin throwers to weaken resistance, before trying to board.

The necessity for manpower to propel the vessel greatly limited the cruising range of such ships, as they had to stop to take on more water and food. Across the world, this was a major limitation to oared warships, whether triremes or the war canoes of, in particular, the Pacific. Furthermore, there were no living or sleeping quarters on ancient ships, which was one reason why they rarely abandoned the coastline and generally beached every night.

Naval warfare played a key role in the fate of the Classical world, with the two crucial battles being Salamis (480 BC) and Actium (31 BC). Salamis arose from the Persian invasion of Greece under Emperor Xerxes I. An earlier invasion, on behalf of Darius I, had been unsuccessful in 490, being defeated at Marathon when the Persian army was charged down by rapidly advancing Athenian infantry.

SALAMIS

In 480, in contrast, a much larger Persian army attacked Greece, supported by a massive fleet of about 1,200 warships. The Greek blocking force was outflanked by the Persian army, a small Spartan rearguard being destroyed at Thermopylae. The Persians then occupied Athens, while the Greek fleet assembled by the nearby island of Salamis. In the face of the larger Persian fleet (about 800 ships to the Greek 300), the Greeks decided to fight the Persians in the narrows of Salamis, rather than in the open water, as they correctly anticipated that this would lessen the Persians' numerical advantage. The Persians indeed found their ships too tightly packed, and their formation and momentum were further

Battles at Sea

480BC BATTLE OF SALAMIS	31BC BATTLE OF ACTIUM	1499 BATTLE OF ZONCHIO	1571 BATTLE OF LEPANTO	1788–90 RUSSO-SWEDISH CONFLICT
The Greek force of just 300 triremes defeat Xerxes' Persian navy with over 800 galleys	Octavian Caesar defeats the combined fleets of Mark Antony and Cleopatra	The Ottomans overcome the Venetians in the first naval battle in history to use cannon on ships	Don John of Austria leads the Holy League to crush the Ottoman fleet led by Ali Pasha	Galleys still used in sea battles in the Gulf of Finland

disrupted by a strong swell. The Greeks attacked when the Persians were clearly in difficulties, and their formation was thrown into confusion. Some ships turned back while others persisted, and this led to further chaos which the Greeks exploited. The Persians finally retreated, having lost over 200 ships to their opponents' 40, and with the Greeks still in command of their position. As a result, Xerxes returned to Anatolia with the remnants of his navy and part of his army. The force he left in Greece under Mardonius was defeated in 479 at Plataea, a land battle.

At Actium, the struggle was between Octavian, later Augustus Caesar, who controlled the western part of the Roman empire, including Italy, and his rival, and former brother-in-law, Mark Antony, who dominated the eastern part. Mark Antony was supported by his wife, the beautiful Cleopatra VII, ruler of Egypt. Actium, on the western coast of Greece, was Mark Antony's anchorage, but it was a poor position because malaria weakened his forces, while their supply routes were endangered by Octavian's nearby troops. This affected morale, leading to desertions among the rowers.

When part of Mark Antony's fleet tried to break out, it was defeated by Octavian's fleet. This led Mark Antony to resolve to break out with the entire fleet. He was seriously outnumbered. The battle showed the necessity of tactical planning. In order to help break through the opposing centre and provide an opportunity for some of the warships, as well as merchantmen carrying their war chest, to escape, Mark Antony had the galleys on his wings move from the centre. He hoped this would lead his opponents to respond, weakening their centre. Indeed, this tactic succeeded, enabling some of his fleet to escape – some 70 warships were saved – but the abandonment of his army was a serious blow, and Octavian was able to press on to conquer Egypt – after both Mark Anthony and Cleopatra committed suicide.

AMPHIBIOUS ATTACKS

Tactics remained similar for centuries – some 1,500 years later the Ottomans had created a fleet to help prevent the Byzantine capital, Constantinople (modern Istanbul), from being relieved when they successfully attacked it in 1453; an earlier Ottoman fleet had been destroyed by the Venetians in the Dardanelles in 1416. The subsequent Ottoman move to the port capital of Constantinople, whose supplies depended on maritime links, led to an increased role for naval concerns in their policy. The Ottoman fleet swiftly became a major force in the Aegean Sea, employed to support amphibious attacks on Venetian bases such as Mitylene (1462) and Negroponte (1470).

The Ottomans subsequently developed their fleet for more distant operations beyond the Aegean and to carry cannon, which they used with effect against the Venetians at the battles of Zonchio in 1499 and 1500.

On galleys, cannon were carried at the front and supplemented the focus on forward axial attack already expressed by the presence of a metal spur in their bow. This spur might damage enemy oars and might be pressed into the hull of an enemy galley if a boarding was attempted. The strengthened and lengthened prow provided the access/boarding ramp into an enemy ship. These spurs could not sink ships as (underwater) rams were intended to do, but the latter was a weapon of the Classical period that had disappeared. Like spurs, cannon were intended to disable the opposing ship as a preparation for boarding.

Galley conflict was not unchanging. Aside from the introduction of cannon, galleys became easier to row in the mid-16th century. The traditional *alla sensile* system, with each oar pulled by one man (with the typical galley having three men on a bench with one oar each), gave way to rowing *a scaloccio*, in which there was one large oar for each bench, and this oar was handled by three to five men. This led to an increase in the number of rowers and also made it possible to mix inexperienced with trained rowers without compromising effectiveness. Experts were long divided about the respective success of the two systems, which is a reminder of the need for caution today when considering the relative strength of military methods. Both systems required large numbers of rowers, which was a major problem with the use of galleys.

Galleys proved their value in the Ottoman victory at Prevesa in 1538 against a Venetian-Habsburg fleet. The Ottomans withdrew their galleys onto the beach, with their forward galley guns facing out to sea, providing a stable gun platform, and, when the Christian fleet retreated, the fresh crews in the Ottoman galleys were able to overhaul and board some of the Christian ships.

BATTLE OF LEPANTO

At Lepanto, the cannon of six Venetian galleases played a particularly significant role in disrupting the Ottoman fleet. Galleases were three-masted lateen-rigged converted merchant galleys which were longer and heavier gunned than ordinary galleys. They also carried firing platforms at poop and prow, and sometimes along their sides as well. Their height also gave them a powerful advantage as they could fire down on opponents. If they were able to crash into the side of, or sweep away the oars of, an enemy galley, the impact of their weight was much larger than that of a normal galley. Lepanto was the last really large-scale galley battle in history, with about 230 ships on each side, and more than 100,000 men involved. Casualty figures vary, but the Ottomans lost about 113 galleys sunk and 130 captured, while the victors lost about 12 galleys.

In the 17th century, sailing ships became increasingly important in Mediterranean warfare. They freed warships from dependence on a network of local bases, carried more cannon than galleys, and were also less susceptible to bad weather.

Galleys remained significant in waters that were too shallow for deep-draught warships, such as the Gulf of Finland, where galleys were key in the Russo-Swedish conflict in the eighteenth century, including in the 1788–90 war. Furthermore, oared boats could give greater flexibility than those that were reliant on sail. Therefore outside Europe, Western warships had only limited value in inshore conditions, as well as in estuaries, deltas and rivers, a situation that was not to change until the introduction of shallow-draught steamships carrying steel artillery in the 19th century. For example, in the 16th century, African coastal vessels, powered by paddles and carrying archers and javelinmen, were able to challenge Portuguese raiders on the West African coast. Although it was difficult for them to storm the larger, high-sided Portuguese ships, they were, nevertheless, too fast and too small to present easy targets for the Portuguese cannon.

South-East Asian rulers responded to the threat posed by European warships not by copying them, but by building bigger armed galleys whose oars gave them inshore manoeuvrability. In the second half of the 16th century, Aceh, Johor, Bantam and Brunei developed substantial fleets of war galleys. In the 18th century, the Illanos of the Sulu Islands deployed large fleets of heavily armed galleys, that were appropriate for inshore operations and able and willing to attack the warships of the Dutch East India Company. Gia-long of Vietnam (1802–20) constructed square-rigged galleys.

This long-term effectiveness of galleys indicates the continuity of military challenges and circumstances and shows the appropriate nature of these vessels for in-shore operations.

Stirrups

'Happiness lies in conquering one's enemies,
in driving them in front of oneself, in taking
their property, in savouring their despair,
in outraging their wives and daughters.'

GENGHIS KHAN

STIRRUPS ARE NOW ASSOCIATED WITH RIDING HORSES, BUT, FOR A LONG TIME THIS WAS NOT THE CASE. The introduction of the stirrup led to a major increase in the capability of horsemen, and was important in increasing the greater effectiveness of cavalry.

The stirrup was developed in Central Asia, the region where the horse had first been domesticated, and was certainly in evidence there by the first century AD. The genesis of the stirrup was a long one. It is possible that the Scythians used leather loops in the fourth century BC, although these may simply have been to help in mounting the horse. These loops were not, therefore, able to provide a better fighting platform. The latter was offered by the use of rigid metal stirrups, which provided stability in motion, helping in both shock action and with firing or throwing projectiles from horseback; in other words assisting both heavy and light cavalry. These actions did not depend on stirrups, but stirrups helped make them more effective. The earliest Chinese figurine with two stirrups probably dates from about AD 322.

Stirrups made it easier for a cavalryman to stay on his horse and so to use the horse as a fighting platform in contact conflict, as with a sword, lance or axe. Such conflict led to shocks that could otherwise unseat a cavalryman. Avoidance of this was taken further with developments in saddlery.

It is important not to exaggerate the impact of stirrups, as the horse had been used effectively in warfare long before they were developed, and many of the features noted in cavalry warfare with stirrups had been anticipated earlier. Stirrups indeed were an improvement, but an incremental one. Furthermore, their use diffused slowly. While stirrups provided a small advantage to cavalry facing other cavalry that lacked them, they made less of a difference to cavalry fighting infantry.

NOMADIC LIFESTYLE

It is essential to focus, in judging the capabilities of steppe forces against less mobile foes, on the effectiveness brought by the techniques of cavalry use which were by-products of a nomadic lifestyle, rather than the technology of the stirrup. Aside from stirrups and saddles, cavalry also benefited from the adoption of more effective edged weapons, and from improvements in horse stocks. This ensured that shock, mobility and firepower were all possible, a lethal combination.

The stirrup contributed to the potency of invading armies both tactically and operationally and was used by the Huns who invaded Europe in the fifth century. The Arabs were particularly effective in their use of cavalry. For example, they brought the bridle and stirrup to North Africa, across which they advanced, capturing Alexandria in 642 and Tripoli in 647. Islam helped to give the series of Arab attacks, there and elsewhere, greater cohesion than those of other 'barbarian' invaders of the established Eurasian civilizations, although the caliphate, and, with it, Islamic unity, was disrupted from the 650s. Under the Umayyad caliphs (661–750), the process of Islamic expansion continued. The remainder of the coast of North Africa was overrun and, in 711, the Berber general Tariq led his men across the Strait of Gibraltar.

The Visigoths, one of the 'barbarian' tribes that had overthrown the Western Roman empire, who afterwards ruled most of Spain and Portugal, were defeated by the Arabs at Rio Barbete, and most of Iberia was conquered. This was not the end of Arab advances: in 720 the Pyrenean mountains were crossed (into modern France), and the city of Narbonne was captured, Toulouse following in 732.

CHARLES MARTEL

However, there were also checks, indicating the extent to which no one form of warmaking was invincible. The Arab advance was stopped at Covadonga in northern Spain in 718, and at Poitiers in France in 732, in the latter case by Charles Martel. Abd ar-Rahman, governor of Spain, had invaded Aquitaine (southwestern France) with considerable success, defeating the local forces, but his advance on Tours was stopped by a Frankish infantry phalanx under Martel. We know very little about the battle, and both armies were probably infantry for the most part. It seems probable that the Franks were harried by Muslim arrows, but, when the Muslims closed to attack, they suffered heavy losses, including their commander, at the hands of their opponents' effective short sword and spear combination. The solidarity and unity of Martel's infantry formation was the deciding factor in the battle. The Muslims fell back, never to repeat their advance so far north.

As a sign, nevertheless, of the range of their power, in 751, near Atlakh on the Talas river near Lake Balkhash, an Arab army under Ziyad bin Salih, governor of Samarkand, defeated a Chinese counterpart under Gao Xianzhi, helping to ensure that the expansion of the Chinese Tang dynasty into western Turkestan was halted and, instead, driving forward a process of Islamicization in Central Asia, the effects of which have lasted to the present day. The battle was decided when an allied contingent of Qarluq Turks defected to Ziyad. Gao's army was badly defeated.

It is also important to note that not all Arab cavalry was light cavalry. For example, the Khurasaniya, on whom the Abbasid caliphs (750–1258) relied, were heavy cavalry, equipped with armour, and armed not with bows, but with curved swords, clubs and axes. During the First Crusade (1097–9), the Crusaders encountered Agulani (probably drawn from Persia) with armour of plates of iron, which also covered the horses: heavier than anything seen in Western Europe.

By then, Western European cavalry was employing the stirrup in mounted shock combat. Long stirrups and a high-backed saddle helped anchor the knight, so that he could sustain the impact of charging with a lance tucked under the right arm, and thus killing or unhorsing rival horsemen. Knights frequently fought dismounted, but the stirrup gave them greater value as cavalry.

Stirrups

First century AD
The stirrup is developed in Central Asia where horses are first domesticated

AD 322
Chinese figurine shows horse with two stirrups

642
Arab cavalry, riding with both bridles and stirrups, capture Alexandria

732
Arabs advance into Europe and capture Toulouse

11th century
Western European cavalry uses stirrups and high-back saddles to anchor knights in mounted shock combat

13th century
The Mongols use their horse-riding skills allied to short stirrups which allow them to stand upright while firing. Under Genghis Khan (c.1160–1227) the vast Mongolian empire is formed

THE MONGOLS

The continued effectiveness of horsemen was shown by the light cavalry of the Mongols who, in the 13th century, overran not only China but also the Abbasid caliphate of Persia. Mongol forces advanced as far as Poland, Hungary, Serbia, Syria, Burma, Java and Japan. They were also the first steppe force to conquer China south of the River Yangzi. Their slaughter of the population of Changzhou discouraged resistance there.

The Mongols used cavalry, specifically mounted archers, to provide both mobility and firepower, and each rider had several horses. They employed short stirrups, which had a major advantage over long stirrups for accurate fire as the rider's torso was free of the horse's jostlings, while his legs acted as shock absorbers. The use of short stirrups required greater ability as a rider as it was easier to be unhorsed, but the use of these stirrups helped provide a steady firing platform, although great skill was needed in order to shoot arrows from the saddle in any direction and also when riding fast. Standing in the stirrups provided more accuracy than bareback riding or the use of cavalry saddles. A Song Chinese general noted of the Mongols, 'It is their custom when they gallop to stand semi-erect in the stirrup rather than to sit down'. Aside from skill in riding, they also benefited from the hardiness of the horses they used.

The Mongols were highly disciplined and adept at cavalry tactics, such as feigned retreats, and at seizing and using the tempo of battle. This was seen in their invasion of Europe in 1241–2. In 1241, a Polish-German army was defeated at Leignitz, while the Hungarians were crushed at Mohi. At Leignitz, the horns of the more numerous Mongol deployment outflanked the Poles, who were hit hard by archers from the flank, while, at Mohi, the Mongols turned the unprotected Hungarian left and, under pressure from front and flank, the vulnerable Hungarians, who had lost tactical flexibility, were defeated with heavy casualties.

Mongol power, nevertheless, did find its bounds. After Liegnitz and Mohi, many European cities fell, including both Buda and Esztergom (Gran), but others, including Olomouc and Brno, successfully resisted, while the Mongols decided it was not worth seriously attacking others, including Split. Furthermore, the Mongol invasion of the Dalmatian Alps was not particularly successful as cavalry could achieve little in the mountainous terrain. Mountain fortresses had also hindered the Mongol advance in southern China. Nevertheless, in 1242, the Mongols only turned back in Europe when news arrived that the Great Khan, Ögödei, had died. Large-scale Mongol attacks on Japan from Korea failed in 1274 and 1281, in large part because of storms.

Once their empire was flung wide across much of Eurasia, the Mongols failed to maintain their cohesion or dynamic. Effectiveness in cavalry warfare could only achieve so much. The 14th century was an age of Mongol decline, in part because of conflict within the ruling élite. Moreover, in China, rebellions against Mongol rule in southern China led to the establishment of the Ming dynasty in 1368 after the Mongol emperor fled from Beijing ahead of the approaching Chinese army. At the other end of the central Asian world, Mamai of the Golden Horde was defeated at Kulikovo in 1380 by an army commanded by Grand Prince Dmitrii Donskoi of Muscovy.

Stirrups remained crucial to cavalry capability, but no other cavalry force was ever to be as effective as the Mongols.

Bows

'The antique Persian taught three useful things,
To draw the bow, to ride and speak the truth.'

DON JUAN, BYRON

BOWS AND ARROWS WERE THE KEY MISSILE WEAPONS UNTIL THE COMING OF FIREARMS, although slings and javelins were also significant. Bows themselves were very varied. For example, over the centuries, there were different models and varieties of the composite bow, evidence of which was seen in Mesopotamia (modern Iraq) in about 2200 BC. Storing compressive and tensile energy by virtue of its construction and shape, allowing it to be smaller than the long bow, the composite bow was a sophisticated piece of engineering that seems to have been invented in different places by various peoples, although there was probably some interchange of ideas. Thus, the Turkish bow was different from the Chinese bow. The composite bow was more effective than the simple bow because its stave of wood was laminated, but its manufacture was labour-intensive. There is no reliable evidence to claim that the composite bow drove out other types of bow; it did not.

The heyday of the bow was in the early Middle Ages when archers were important both as infantry and as cavalry. In the seventh century AD, a major Arab expansion included the defeat of Byzantium (the Eastern Roman empire) and the overthrow of Sassanian Persia (Iran). The Byzantines were defeated at Yarmuk and the Persians at Al Qadisiya, both probably in 636, in part due to the impact of the Arab archers, although the nature of the surviving sources is such that it is difficult to make comments about the army structure or size, weapons and tactics. The Arabs pressed on to the east, to capture Herat and Merv in 650.

SALADIN'S MOUNTED ARCHERS

Mounted horse archers of Central Asian origin after this were frequently the key soldiers in the Near East and proved the most appropriate answer to the dangers facing the Islamic world in the 12th and 13th centuries: the Crusades and especially the Mongols. These archers repeatedly proved effective in battle. In 1071, at Manzikert, helped by divisions and poor command among their opponents, the Seljuk Turks crushed the Byzantine army, following up by capturing much of Anatolia. In 1187, Saladin destroyed the Crusaders at Hattin. In this battle, William of Tyre captures the harassing role of Saladin's mounted archers. He wrote of the Crusaders:

> They left the Springs of Saffuriya to go to the relief of [the besieged fortress of] Tiberias. As soon as they had left the water behind, Saladin came before them and ordered his skirmishers to harass them from morning to midday. The heat was so great that they could not go on so as to reach water. The king [Guy of Jerusalem] and all his men were too spread out and did not know what to do.

The Mamluks also had very effective archers, who defeated the Mongols in Palestine and Syria at Ain Jalut (1260) and near Homs (1291). The Mamluks used 'slower-shooting': rapidly firing up to five arrows held in the hand at the same time, whereas the Mongols lacked this high-speed archery. The Mamluks also proved effective against the Crusaders, taking their last outpost in Palestine, the fortress-city of Acre, in 1291.

Light cavalry armed with bows were also vital to the forces of Timur the Lame (1336–1405; later called Tamerlane). Having gained control of Transoxiana, where his capital was at Samarkand, Timur conquered Herat in 1381, before turning on Persia. He also attacked the Mongol Khanate of the Golden Horde in 1388–91 and 1395, sacking their capital New Sarai on the River Volga in the latter year after he had won a major victory in the Terek Valley. In 1398, Timur advanced across the mountains of the Hindu Kush into India where he sacked Delhi. In turn, Baghdad and Damascus fell in 1401, while the Ottoman (Turkish) army was defeated at Ankara in 1402. In this battle, Timur benefited from numerical superiority – and also by depriving the Ottomans of water by destroying wells and diverting a creek. The battle was settled when the Kara Tatars on the Ottoman left switched sides.

Timur's forces indicated the extent to which light cavalry forces were not unsophisticated 'nomadic hordes'. Instead, they were highly organized and well-disciplined, and this helped him execute such methods as rapidly changing the direction of march. As a result, it was not so much light cavalry itself that was crucial, but the way it was employed. Indeed, some of Timur's opponents, such as the khanate of the Golden Horde, also used light cavalry. Timur also had about 30 elephants, but these were not fundamental to his victories.

> Timur stayed in Anatolia for a year, inflicting terrible devastation and restoring lands conquered by the Ottomans to their previous emirs. He also advanced to the Aegean, storming the stronghold of the crusading Knights of St John at Smyrna.

A PYRAMID OF SKULLS

Timur raised supplies on the march, but, rather than simple devastation, efforts were made to use existing structures by levying tribute on the basis of lists of businesses and tax registers. In captured cities, there was a systematic attempt to seize goods, not disorganized plunder. Indeed, Timur preferred to persuade cities to surrender and then pay ransom. He only stormed cities when this failed, but, when he did so, he used brutal force towards the inhabitants in order to deter other cities from resistance, most vividly by erecting pyramids from the skulls of the slaughtered – possibly 70,000 people when a rising at Isfahan in Persia was suppressed in 1388. In these cases, the few who were not killed were treated as slaves and marched to Transoxiana, many dying on the way.

After his victory at Ankara, Timur stayed in Anatolia for a year, inflicting terrible devastation and restoring lands conquered by the Ottomans to their previous emirs. He also advanced to the Aegean, storming the stronghold of the crusading Knights of St John at Smyrna in 1402. Timur then marched east, with the Ottomans, now divided by rivalry between Bayazid's competing sons and no longer a threat to his position. Instead, Timur's support was sought by numerous rulers, including the sultan of Cairo and Christian European powers. At the time of his death in 1405, Timur was planning to invade China.

While emphasizing the role of mounted archers, it is appropriate also to comment on infantry archers. For example, in 750 at Tell Kushaf in Iraq (Battle of the Zab), where Umayyad power was overcome by the outnumbered Abbasids, the latter dismounted forming a spear-wall from behind which archers fired. Disunity ruined the Umayyad response, underlining the importance of cohesion. The role of infantry tactics in the battle was characteristic of much early Islamic warfare, and serves as a reminder of the extent to which it should not be seen solely in terms of cavalry.

Archers were also of value to other powers. On the Ugra river in 1480, Khan Akhmet of the Golden Horde was defeated not so much by the arquebusiers of Ivan III of Muscovy as by his archers. In the early sixteenth century, the Ottoman Turks showed that horse archers were ineffective in the face of gunpowder weaponry; but this was far less the case in South and Central Asia. Furthermore, the Ottomans had made considerable use of the composite recurved bow and their bowmen could fire six arrows a minute.

THE MUGHALS

The Mughal ability to control the supply of war horses in India ensured that they dominated mounted archery in the sub-continent in the 16th century, and this archery played a major role in a relative decline in the importance of elephants, which had previously been dominant in India. In 1556, at the Second Battle of Panipat, the Mughals defeated a much larger insurrectionary force thanks to their use of mounted archers. In China, archers were also used by Zheng Chenggong in his attack on Nanjing in 1659, and they proved more effective than his musketeers. Archers were used by the Chinese throughout the 18th century. The Baluchis (of modern Pakistan) used the bow in the 17th century, but acquired firearms in the early 1700s. The Lezgis in the Caucasus had made a similar shift by the 1720s.

Archers were widely used in Africa in the 16th and 17th centuries. In 1530, the *jihadis* in East Africa who attacked Ethiopia bought cannon from the Ottoman Turks, but also hired a force of Arabian archers. The forces in the successful *jihad* launched in northern Nigeria in 1804 initially had no firearms and were essentially mobile infantry, principally archers, able to use their firepower to defeat the cavalry of the established powers, as at the Battle of Tabkin Kwotto (1804).

Bows continued to play a major role in conflict until the 19th century, although, by then, they were definitely outclassed. In the 1830s and 1840s, in West Africa, King Gezo of Dahomey, whose troops were armed with firearms, conquered the Mahi to the northeast, who had successfully resisted Dahomian attack in the 1750s. Mahi warfare was based on bows, as Dahomey, on the Atlantic coast, prevented the sale of European firearms to peoples in the African interior. Elsewhere, those armed with bows also fell victim to firearm forces or adopted the gun. Nevertheless, like the spear and the sword, the bow had been a crucial weapon for much of recorded history.

Archers were widely used in Africa in the 16th and 17th centuries. In 1530, the *jihadis* in East Africa who attacked Ethiopia bought cannon from the Ottoman Turks, but also hired a force of Arabian archers. The forces in the successful *jihad* launched in northern Nigeria in 1804 initially had no firearms and were essentially mobile infantry, principally archers, able to use their firepower to defeat the cavalry of the established powers...

Longboats

'The heroes, the warriors on their eagerly-
sought adventure, pushed off the vessel of
braced timbers. Then with foam at its prow,
like a bird, it floated over the billowing waves,
urged onwards by the wind... the curved prow
had journeyed so far that the voyagers saw the
land, the sea-cliffs, glisten...'

BEOWULF

WITH THEIR SAILS, STEPPED MASTS, TRUE KEELS AND STEERING RUDDERS, Viking longboats, although shallow and open, were effective ocean-going ships. As a result, they did not need to hug the coasts, as other ships had tended to do, but, instead, could make bold sea crossings. This made it difficult to prepare against them. Clinker-built, streamlined and low, the tactical flexibility of the longboats was enhanced by the provision of a prow at either end. This made speedy relaunching easier. Aside from being able to sail, they were also able, thanks to their shallow draught, to row in coastal waters and up rivers, even if there were only three feet of water.

Thus, the key advantage offered by the longboat was one of mobility, rather than any particular value in battle, although, as far as the latter was concerned, the longship was also fit for purpose. This mobility and flexibility were key advantages but discussion of naval capability has tended to focus on ships that lacked such flexibility, particularly vessels suited to deep-sea operations. However, the powerful Ships of the Line that engaged in operations across the globe in the 18th century were unable to take part in inshore tasks without serious problems, let alone go up river systems. In contrast, longboats were far more multi-purpose. In Ireland, for example, where the wealthy monasteries attracted attack, the numerous rivers and lakes facilitated Viking movement.

THE VIKINGS

The Vikings profoundly influenced northern Europe. They were adventurers, originating in Scandinavia, who ranged far and wide across the North and Baltic Seas, travelling vast distances in their sailing ships (in contrast to the Anglo-Saxon ships which were probably only coastal-huggers, although bolder claims have been advanced); and using oars where needed.

The Vikings moved down through Russia to Byzantium; across the Atlantic Ocean, to establish farming communities in Iceland, Greenland and, for a short time, Vinland on the coast of Newfoundland; and colonized many territories in Britain, including the Orkneys, Shetlands, Hebrides, and Isle of Man, as well as Normandy in France. The limited availability of land for colonization in Scandinavia encouraged their expansionism.

To the west, the attraction of the British Isles for raiding and settlement led to attacks. Danish ships were first recorded in English waters in AD 789, and the great Northumbrian monasteries of Lindisfarne and Jarrow were brutally sacked in 793–4. From the 830s, Viking raids on the British Isles and France increased in intensity. Soon the Vikings came to conquer, not to plunder. They came with larger forces that overwintered, and also developed permanent coastal bases. In England, the Danish Great Army overran East Anglia in 865, Yorkshire in 866–7, and Mercia in 874, before being stopped by Alfred, king of Wessex, at the Battle of Edington in 878.

The Norwegians made a major impact on Ireland. They were first recorded in 795 when they sacked Lambay Island off the east coast. From the 840s, their military presence became stronger with larger forces that overwintered in Ireland and developed permanent coastal bases. The first, Dublin, established in 841, was followed by Limerick, Wexford, Waterford and Cork. From such bases, the Norwegians, using their longboats, dominated the Irish Sea and its trade, and intervened on the west coast of England, where they invaded the Wirral from Dublin in 902.

TRADE AND PLUNDER

The Norwegians from the 850s also began pressing on the north Welsh coast, especially the vulnerable island of Anglesey. They defeated Rhodri Mawr in Anglesey in 867. The Vikings also plundered the royal seat of the major Welsh kingdom of Gwynedd at Aberffraw in 968. However, the legacy of Viking place names, such as Anglesey and Swansea (Sweynsey) in Wales, suggests that the Vikings came to trade and settle as well as to devastate.

Evidence of Swedish activity east of the Baltic in the ninth and tenth centuries is also conspicuous. The large numbers of coins from north Russia found in Scandinavia may indicate trade, plunder, tribute or the payment of mercenaries. Islamic texts refer to the 'Rus' as traders who were settled in Kiev as early as the middle of the ninth century. They also held bases at Novgorod and Staraya Ladoga.

On land, however, the Vikings lacked many of the military advantages they had on the water. No better armed than their opponents, the Vikings were not particularly effective in battle, in which their standard deployment was a shield wall. Having invaded Wessex, the Danish Great Army, for example, had turned to France, where it did much damage, but failed in the siege of Paris in 885–6. The fortified bridges built in France under Charles the Bald (r. *c.* 843–77), in order to obstruct Viking passage up rivers, were successful in stopping them.

Moreover, having been defeated by Alfred, and forced to accept territorial limits, the Danes were driven back from England. Alfred's heirs, Edward the Elder, Athelstan and Edmund, brought most of the area of modern England under their authority. Edward overran the Danish bases in eastern Mercia, conquered East Anglia, and built forts in the northwest Midlands, including Manchester (919) to limit the danger of attack from the Norwegian kingdom of Dublin. In 927, Athelstan captured York, the Danish capital in northern England.

Viking attacks on Britain

789
Danish ships first recorded in English waters

793-4
Lindisfarne and Jarrow monasteries are brutally sacked

865
Danish Great Army conquers East Anglia

866–7
Danes overrun Yorkshire

869
Vikings kill King Edmund of East Anglia by shooting him full of arrows, 'like a hedgehog'

874
Danes conquer Mercia

878
Battle of Edington in Wiltshire. The Danes are defeated by King Alfred of Wessex

886
Alfred frees London from Danish occupation and makes a treaty with the Viking leader, Guthrum

IRISH RESISTANCE

In Ireland, as in England, resistance to Viking invasion brought a measure of political consolidation. Máel Sechnaill II, High King of Ireland, defeated the Dublin forces at the battle of Tara (980), and Brian Boru defeated the Norse of Limerick at Sulcoit (*c.* 968) and routed his Irish rival, Máel Mórda, King of Leinster, and the latter's Viking allies from Dublin, Orkney and Scandinavia, at Clontarf (1014). This battle decisively weakened the Vikings in Ireland.

However, there were also renewed Viking advances. In 911, Charles the Simple, ruler of France, ceded Rouen and the lower Seine valley to a Viking group led by Rollo, who extended their authority to the west and took over the entire area now known as Normandy. They were later known as Normans, and were eventually to conquer England (in 1066), as well as southern Italy and Sicily.

Meanwhile, a fresh wave of attacks on the British Isles in the late tenth and early 11th century, under King Swein of Denmark, led to the Danish conquest of England. It was ineptly defended by Aethelred the Unready, but he may have faced Danish armies that were at times larger than those that attacked Alfred's Wessex. Swein's son, Cnut, became ruler of England (in 1016), Denmark (1019), and, after its conquest in the 1020s, Norway. In 1031, Cnut advanced to the River Tay and received the submission of Malcolm II of Scotland. After his death in 1035, however, his empire dissolved. The next Viking attack on Britain, that by King Harald Hadrada of Norway in 1066, was initially successful, but was speedily crushed by King Harold of England at the Battle of Stamford Bridge outside York.

In 1263, at the battle of Largs, the amphibious force of Hakon IV of Norway was defeated by the Scots, which led to Scotland gaining the Western Isles by treaty in 1266. The Viking political axis along the northern and western margins of Scotland was gravely weakened as a result of both battle and treaty.

COMMERCIAL DEVELOPMENT

Alongside the emphasis on warfare, it is important to note that the Vikings helped develop important trading links. Their enterprise opened up the North Sea, and contributed substantially to the later mercantile and commercial development of northern Europe. As already mentioned, the Norwegians reached Iceland in about 860, Greenland in 986, and Newfoundland and the coast of North America in about 1000, but were unable to develop this route. Instead, the settlements in Greenland and America were abandoned. In Greenland, the Vikings fell victim to its remoteness and the impact of a deteriorating climate on an already harsh environment and, probably, also the resistance of the local Eskimos.

The Viking tradition of longship building continued for centuries. For example, it was important in the islands off west Scotland until the late 16th century, providing mobility and a continued capacity for raiding. In 1533, the English attempt to impose their power by using new naval technology failed when the *Mary Willoughby* was captured by longships off the Shetlands. The use of Viking-style longships by the Western Isle men, notably the MacDonalds, enabled them to transport troops from the Western Isles of Scotland into Ulster, using Lough Foyle as their landing area, throughout the 16th century. Attempts to intercept the fleets by English sailing ships were unsuccessful, as were efforts to penetrate into the lochs of the Mull of Kintyre to attack the vessels at source. Only the occupation of Lough Foyle by the English finally solved the problem.

By then, the Scandinavian kingdoms – Denmark (which also ruled Norway and Iceland), and Sweden (which also ruled Finland) – had navies of a conventional European type. The shallowness of parts of the Baltic and its limited tides were to encourage the additional use of shallow-draught ships, especially by the Swedes in the Gulf of Finland in the 18th century, but they were very different from longboats. Yet these boats had helped the Vikings to play a major role in the development of northern Europe.

Castles

'For the fortifications called castles by the French were scarcely known in the English provinces.' (Explaining the weakness of the English resistance despite their fighting prowess.)

ECCLESIASTICAL HISTORY, ORDERIC VITALIS, 1075 – c. 1142
ENGLISH-BORN NORMAN monk and chronicler

ARLY CASTLES WERE GENERALLY SIMPLE AFFAIRS OF EARTH AND
TIMBER CONSTRUCTION thrown up in a hurry, although they still
required many man days to construct. Timber-built forms were often
of either motte (a mound of earth)-and-bailey, or ringwork (enclosure)
form, both of which had a long currency up to the 13th century. Unlike hill
forts, which had been numerous in the Iron Age, castles were inhabited,
rather than places of refuge. However it is important to note that ancient
and Roman fortifications were also extensive.

Not all early castles were built of timber and earthwork. There was also construction in
stone, which, crucially, was not vulnerable to fire and therefore was more resistant to attack.
However, cost and the shortage of skilled masons affected the attraction of construction in
stone but, nevertheless, with time, major castles were fortified in stone.

Castles were particularly effective as part of a combined military system. European
military history in the 11th to 13th centuries in part centred on how the Frankish
development of knights, castles and siege techniques enabled the rulers who employed them
to extend their power, both against domestic opponents and on their frontiers. The latter
forced the rulers of more peripheral regions, such as Scotland, Prussia and pagan Lithuania,
to build forts and castles.

ROYAL GOVERNMENT
Castles were designed to repel external attack, but also to ensure domestic control. In
England, William I (William the Conqueror), king from 1066 to 1087, and his royal
successors maintained castles in the shire towns (as well as the Tower of London) as part of
their framework of royal government. Surviving examples of castles provide ample
illustrations of how they literally towered over and commanded the surviving countryside.
Alongside their defensive purpose, castles had a key function as fortified residences, the
importance of which is easily overlooked.

On the frontiers of Christendom, castles were built to consolidate Christian
expansion. This was the case in Spain and Portugal during the lengthy wars with the Moors,
and also in the Baltic lands where the 'Northern Crusades' were being staged. Castle
construction was particularly important in the Middle East during the Crusades. Castles
there were a potent display of power and a way to derive maximum advantage from a
relatively small number of men, and, for these reasons, many were built by the Crusaders.
With the development of the concentric plan from the late 1160s (Belvoir 1168–70, is the
first concentric castle), castle design in the Holy Land advanced more rapidly than in
Western Europe, but the challenge the Crusaders faced was more serious. Indeed, better
Muslim siege techniques led the Crusaders to adopt the new plan.

KRAK DES CHEVALIERS
For example, the outer walls of Krak (or Crac) des Chevaliers were strengthened by semi-
circular angle towers. The multiple character and interlocking strength of the defences there
ensured that, once the outer walls had been penetrated, it was still necessary to advance
along overlooked corridors and ramps. Krak itself indicated the long and varied history of
specific military sites. Originally, Krak was built on by the Arabs to guard the route from
Tripoli to Hamah – most castles would occupy a strategic location. The Crusaders developed
the fortifications under the Count of Tripoli, but the castle that survives was totally rebuilt

by the Hospitaller military order after they obtained it from the count in 1142.

Krak successfully held out against the Kurdish general Saladin in 1188, unusually so because the previous year the army of the Crusader kingdom of Jerusalem was defeated by Saladin at Hattin. The destruction of the field army left most of the fortress garrisons in too weak a position to hold out and Jerusalem surrendered after its walls were undermined and breached. Acre also fell. The fate of the fortresses indicated some of the weaknesses of castles, specifically the danger of denuding them of troops in order to create a field army, and also the psychological effects of the defeat of such an army on remaining garrisons.

Mamluk success at siegecraft was crucial to the eventual failure of the Crusader kingdoms. The Mamluks captured Haifa in 1265, Krak in 1271, Tripoli in 1289 and, at last, Acre. It fell in 1291 to a determined assault, after the defences had been weakened by mining and by stone-throwing siege engines.

Castles were also used to try to consolidate new frontiers within Christendom. The English conquest of North Wales was marked by extensive fortress construction. Castles were built in Wales both by the English Crown and by aristocrats seeking to expand their power. William I was probably responsible for building Cardiff Castle when he visited Wales in 1081. One of his lords, William fitz Osbern, had already built a stone castle, the first in Wales, at Chepstow in 1067–71, and used it as a base for expansion. Another, Roger of Montgomery, founded a castle at Montgomery, and advanced from there into central Wales. He was also responsible for pressure further north from the castle at Oswestry.

Castles provided refuge from Welsh attack, but could also be easily bypassed. Thus, fortress construction did not fully stabilize Norman control. Instead, the Normans had to rely on light forces of their own to pursue Welsh raiders. Lightly armed cavalrymen, linked to castles, were one remedy.

> Caernarfon, a fortress-palace, Conwy, Harlech and Beaumaris – built in Wales after Edward I's campaign of 1282 – were all coastal castles that could be supplied by sea, where the English were dominant.

COASTAL CASTLES

The major new fortresses – Caernarfon, a fortress-palace, Conwy, Harlech and Beaumaris – built in Wales after Edward I's campaign of 1282 – were all coastal castles that could be supplied by sea, where the English were dominant. Most of the heavy building materials for their construction were brought the same way. The construction of these massive stone-built works was a formidable undertaking, costing at least £93,000 and using thousands of conscripted English workers. The castles, nevertheless, provided their value in subsequent operations, anchoring the English presence in the face of successive rebellions. English expansionism in Ireland was also marked by the construction of castles, such as Carrickfergus, Dundalk, Coleraine, Trim, Athlone and Kildare.

Castles came to seem redundant from the 15th century in the face of cannon, although new works were still constructed. For example, in England, Ralph, 3rd Lord Cromwell, who in 1433–43 was lord treasurer to Henry VI, built a castle at Tattershall. Its Great Tower was built of brick, then a relatively novel and therefore prestigious building material. The castle was probably as much for show as for defence, but the combination of the tower, moats and gatehouses with Cromwell's large retinue would have made it a formidable structure. Less prominent families also built castles. One at Oxwich, a possession of the Mansels, was first specifically mentioned in 1459.

Castles, however, were by the 17th century frequently disused. Beaumaris in Wales was described as 'utterly decayed' in 1609. A 1627 survey of Conwy revealed that it was in a poor state. The castles were replaced by new-style fortifications, many adopting the style of the Trace Italienne. Nevertheless, medieval fortifications came into their own again during the English (really British) Civil War (1642–6), for example at Chester.

CASTLE KEEP TO STATELY HOME

In part, the redundancy of castles was a response to the increasingly demilitarized nature of society. The shift from castle keep to stately home was symptomatic of an apparently more peaceful society, as well as a result of the heavy costs of castle building. In 1762, Elizabeth Montagu could reflect 'a virtuoso or a dilletanti may stand as secure in these times behind his Chinese rail as the knight on his battlements in former days'. Town walls as well as castles fell into ruin. In England, a survey in 1597 found that Melbourne Castle was being used as a pound for trespassing cattle, and it was demolished for stone in the 1610s. John Speed described Northampton Castle in 1610, 'gaping chinks do daily threaten the downfall of her walls'. When King James I visited Warkworth Castle in 1617, he found sheep and goats in most of the rooms.

In Wales, many castles were abandoned, and fell into disrepair and ruin, while others were enhanced not with fortifications but with comfortable and splendid internal 'spaces', particularly long galleries, as at Raglan, Powis and Carew.

Politics also played a role. Defeated nobles and towns had their fortifications destroyed or weakened. Thus, the Civil War in Britain led, in Wales, to the demolition of Raglan's Great Tower, a potent symbol of the fall of aristocratic power. Other castles 'slighted' included, in Wales, Abergavenny, Aberystwyth, Flint, Laugharne, Montgomery, Pembroke, Rhuddlan and Ruthin. In England, there was also extensive slighting. For example, at Kenilworth, the north side of the keep was demolished and parts of the outer curtain wall were destroyed. Corfe, Dunster and Winchester castles were also among those slighted.

Castles continued to play a role in European conflict thereafter, that at Carlisle, for example, falling to siege twice during the Jacobite rising of 1745, but they were very vulnerable to cannon.

There were also non-Christian traditions of castle building. There was an important one in Japan, and this responded to gunpowder by combining thick stone walls with hilltops of solid stone. About 60 castles built between 1580 and 1630 survive in Japan. Furthermore, major castles were built by Arab rulers, for example the citadel of Aleppo. In China, the emphasis was on city walls, and they were relatively low, extremely thick and made of packed earth, rather than the brittle stone which made European fortresses vulnerable. Chinese earth fortifications were to be surprisingly effective against British warships in 1859 as they absorbed the British shot.

Castles throughout the world still remind us of one of the most impressive and dramatic demonstrations of past military systems.

Longbows

'The English archers then advanced one step forward, and shot their arrows with such force and quickness, that it seemed as if it snowed.'

SIR JOHN FROISSART ON THE BATTLE OF CRÉCY

PROFICIENCY WITH THE LONGBOW WAS CRITICAL TO ENGLISH MILITARY SUCCESS IN THE 14TH CENTURY, and helped define the Hundred Years War between the kings of England and France. This proficiency was also an aspect of the more general rise in infantry capability in the 'mid'-Middle Ages. Knights anyway had always been of limited tactical value in capturing or holding fortresses. Furthermore, the infantry of the Lombard League in northern Italy in the 12th century, and of Flanders and the Swiss cantons in the early 14th century, posed a formidable challenge to cavalry forces. At Legnano, in 1176, Frederick Barbarossa, the Holy Roman Emperor, was defeated by the League and its effective spearmen, while at Courtrai, in 1302, poorly deployed French knights were crushed by Flemish militia with heavy casualties.

So also in the British Isles, where the Scots under William Wallace defeated the English at Stirling in 1297. However, when Edward I of England, in turn, triumphed at Falkirk in 1298, the Scottish pikemen, massed in tightly packed schiltroms, although they were able to defy the English cavalry, were broken by Edward's archers. At Bannockburn in 1314, on the other hand, Scottish pikemen on well-chosen ground routed the English cavalry. We have no definite numbers, but the English handled their archers very badly in this battle.

Nevertheless, archers showed how infantry could defeat both opposing cavalry and infantry. The English were to develop this aspect of their armies in the 14th century, deliberately seeking battle and heavily defeating the Scots at Halidon Hill (1333), and the French at Crécy (1346), Poitiers (1356) and Agincourt (1415). In each case, the English were also brilliantly led, although, as more generally in the period, battle descriptions are frequently contradictory, which makes it difficult to supply precise accounts, let alone reasons for victory. The propensity of the French to launch ill-considered and poorly executed attacks was as important as the defensive capacity of the English (not least the superiority of their archers) in the latter's victories. This compensated for the relatively modest size of the English forces. Edward III probably landed in France with 15,000 troops in 1346, and Henry V with 11–12,000 in 1415.

> **Longbows** were traditionally made out of a single piece of yew wood and had an average draw weight of 100 pounds. A skilled craftsman can make one in a few hours. The bows were two metres (six feet) long and their arrows one metre (three feet) long. These were carved from ash, oak or birch. An experienced archer could fire an arrow every five seconds.

HUNDRED YEARS WAR

The end of the male line of Philip IV led Edward to challenge his nephew, Philip VI (r. 1328–50), the first of the house of Valois, for the throne. Each was pursuing a claim on the French throne that made what became known as the Hundred Years War such an intractable struggle. There were spectacular English successes in the first part of the war, especially at Crécy. There the combat effectiveness of the English longbowmen emerged clearly. The French were supported in this battle by Genoese crossbowmen, but they fired more slowly, so that a skilled longbowman could fire three or four times as fast. The bowstrings of the Genoese crossbows had also been affected by rain. The English fought dismounted with the archers on the flank. The French were poorly commanded, certainly in comparison with the leadership provided by Edward III. The French suffered maybe up to 16,000–17,000 casualties, and the English far fewer.

This stage of the war was ended by the Peace of Brétigny of 1360, by which Edward renounced his claim to the French throne, a realistic decision, as well as to Normandy and Anjou, but was recognized as duke of the whole of Aquitaine (southwest France), as well as ruler of the fortress of Calais, which had been captured in 1347.

This, however, was not a stable settlement. War was resumed in 1369 as a result of French encouragement of opposition to Edward in Aquitaine. Edward reasserted his claim to the French throne, but the war went badly both in Aquitaine and at sea and, by the time of the Truce of Bruges (1375), England held little more than Bayonne, Bordeaux and Calais. There was renewed English failure under Edward's successor, Richard II, but success under Henry V.

BATTLE OF AGINCOURT

Invading Normandy in 1415, Henry captured the port of Harfleur and smashed a French army at Agincourt. The French force was considerably larger, possibly 20,000 to 8,000 English. The English force was deployed between two woods, with dismounted men-at-arms and archers, who also had planted stakes in the earth in front of them in order to block the French advance. The French launched cavalry charges on the English archers on the flanks, but were held back by arrow fire. The woods limited the French room for manoeuvre, and this reduced the value of their superior numbers. The battle became a mêlée, and the French were pressed forward by their numbers and, like the Romans against Hannibal of Carthage at Cannae in 216 BC, were unable to fight effectively. The French suffered very heavy casualties. The English were also helped by heavy rain the previous night, which made the ground unsuitable for the French cavalry. In addition the French were also hit by poor leadership, and certainly had nothing to compare with the effective and bold leadership of Henry V.

Between 1417 and 1419, Henry conquered Normandy, in part thanks to his use of siege artillery, but more due to the weakness of the divided French. He also laid claim to the French throne. Henry's success led the weak Charles VI of France (r. 1380–1422) to betroth his daughter Catherine to Henry and to accept the latter as regent during his reign. Henry and his heirs were to inherit France on Charles's death. Charles VI's son, Charles, however, refused to accept the treaty and Henry had to fight on. In August 1422 he died, possibly of dysentery, while besieging Meaux. Henry left France south of the River Loire unconquered and owing allegiance to Charles VII (r. 1422–61), and the war stretched on.

Longbowmen lacked the operational and tactical flexibility of Central Asian archers because, although they were sometimes mounted for movement on campaign, they did not have the mobility of their Central Asian counterparts. Furthermore, longbowmen could not fire from the saddle and tended to fight on the defensive. This ensured that they depended on being attacked; a general problem with infantry forces in the period. Edward III sought to overcome this by inflicting terrible devastation on rural France in order to provoke a French attack.

This was not, however, the major limitation of the English armies. Instead, the problem was that triumph in battle could only achieve so much, and the English found it impossible to turn victory into a permanent settlement in Scotland, Ireland and France. Crucially, political support proved elusive in all three, while there were also military difficulties. When, for example, the English invaded Scotland, the defenders could avoid battle and concentrate on harrying the English force and denying it supplies, a policy that thwarted Edward III's invasion of Lothian in 1356. In addition, it would have been staggeringly expensive to maintain a large number of English garrisons, and, as Scotland itself would not have been rich enough to be made to fund its own occupation, the cost would have fallen on England.

Similarly, in Ireland, a skilful combination of archers and cavalry provided a valuable military advantage for England. However, Ireland was too distant for a monarchy based in southern England to deploy its resources adequately. Furthermore, because it was politically decentralized, conquest was necessarily piecemeal and slow. By the 1320s, a Gaelic resurgence had compromised English chances of conquest and also ended the surplus that the Irish Exchequer had contributed to the English Exchequer.

THE CROSSBOW

The desirable specifications of weaponry were rarely in concert, and bows were no exception. Crossbows, which required far less training and physical strength than longbows, gained more penetrative power in the 15th century as a result of the development of steel bows. However, as a reminder of the complex trade-off of the characteristics of weapons, these bows were expensive, as tempered spring steel was a costly material to make and one that required much skill.

This was not the case with longbows. They were easier and cheaper to make, but they required lengthy training and considerable expertise for proficient use. In addition, their specifications were never improved as were those of crossbows.

Like longbows, crossbows demonstrated their value in defensive positions. This was seen in the Hussite war in Bohemia (the Czech Republic) in 1420–31. The Hussites held off the attacks by the Emperor Sigismund thanks to excellent leadership and innovative infantry tactics, including the use of fortified wagons to create effective defensive boundaries. The wagon fortresses – *wagenburgen* – were defended with crossbows and also handguns and cannon, the latter signs of a receptiveness to new armaments. The Hussite armies were able to inflict serious defeats, such as Domažlice (1431), on their attackers.

In the 16th century, firearms replaced bows as their potential was grasped. Nevertheless, longbows continued to be important in Britain. The English army that defeated the Scots at Flodden (1513), inflicting heavy casualties, including James IV of Scotland, contained archers, not arquebusiers. Similarly, at Pinkie (1547), an English army, this time invading Scotland, again inflicted a major defeat on the Scots: the Scottish army, again principally pikemen, was badly battered by English cannon and archery. Longbowmen remained the chief deliverers of missiles of the Elizabethan militia into the mid-1580s, despite the government's determination, from the 1560s, to re-arm garrisons and key trained bands with the caliver – an early gun – and, increasingly, with the musket.

In about 1660, Sieur de La Mire proposed the use of archers also armed with pikes for the French army. It was claimed that archers could fire five times as fast as musketeers, inflict more serious injuries, carry more shots, bear harsher weather conditions, and cost less. By then, however, European interest in archery was a curiosity.

Swiss Pikes

'Trail'st thou the puissant pike?'

HENRY V, ACT IV i

WILLIAM SHAKESPEARE

PIKEMEN, NOW ONLY SEEN IN CEREMONIAL USE, DOMINATED THE EUROPEAN BATTLEFIELD at the close of the 15th century, inflicting serious defeats on heavy cavalry.

The use of pikes had a lengthy genesis in the later medieval period. It is possible to note continuities with the spears and halberds used by the Flemings, as at Courtrai (1302), and the Scots, as at Bannockburn (1314), and also with the halberds – a combined spear and battleaxe – employed by the Swiss in the 14th century. The halberd was shorter than the pike, but the tactics used were similar. Pikes were originally utilized as a defensive weapon, but the particular value of pikemen related not only to their ability to sustain a defence, but also to their potential for effective attack. Useless as an individual weapon, because it was heavy and inflexible, and could not provide all-round protection, the pike was devastating *en masse* in disciplined formations. This led to a need for trained, professional infantry, far superior to most medieval levies. The operation of a unit of pikemen required concerted actions that rested on drill. As with musketry, methodical training was required.

The Swiss gained great prestige as pikemen by their defeat of the combined armed forces of Charles the Bold of Burgundy which had especially strong cavalry units at the successive Battles of Grandson (1476), and Murten (1476). These victories were not only due to their pikemen, but also to Charles's poor leadership, and from the cavalry which the Swiss were provided with by allies, as they had none of their own. This ensured the harrying of the retreating Burgundians which helped to make Murten a sweeping victory. Charles the Bold was also defeated at Nancy in 1477, but there were few Swiss there and the battle was largely fought by Lorrainers and Germans.

As a result of these victories, the Swiss Confederation made significant gains. These victories led other powers to hire (France) or emulate (several German rulers) the Swiss in establishing pike forces, with the German rulers developing *landsknecht* units. At Guinegate (1479), the Holy Roman Emperor Maximilian used pikemen to beat a French army strong in firepower.

In defence, pikemen could give protection against cavalry, as the pike outreached the lance. Furthermore, pikemen provided musketeers with the defensive strength otherwise offered by walls or field entrenchments. Other infantry weapons were less able to do so.

THE ITALIAN WARS

Pikemen subsequently played a prominent role in the lengthy struggle to dominate Italy during the Italian Wars (1494–1559), in which, aside from struggles between Italian rulers, France, Spain and the Habsburgs all vied to dominate Western Europe. The Swiss were involved because they had territorial interests of their own in northern Italy, but, more generally, as mercenaries. Their skill and the limited size of the regular forces of the major powers ensured that the Swiss mercenaries had a ready market. The Swiss also obliged the Emperor Maximilian to acknowledge their independence.

The Swiss mercenaries played a prominent role in a number of major battles. At Novara (1513), advancing Swiss pikemen competed with artillery, which inflicted heavy casualties until the experienced pikemen overran the French position: left without protection, the French musketeers were routed.

Yet the Battle of Marignano in 1515 revealed that different conclusions could be drawn about the effectiveness of particular arms. On the first day of the battle, the French cavalry played a major role in resisting the Swiss who were also held by Francis's pikemen. The next day, the attacking Swiss suffered badly from the French cannon, because, on this occasion, the French fought from an entrenched position. The pikemen also suffered fire from crossbows and arquebuses, and were attacked in the flank by French cavalry.

Far from conferring a long-term advantage, the Italian Wars revealed the Swiss formation of large and dense blocs of pikemen to be vulnerable and their tactics to be rigid. To maintain their mobility, the Swiss pikemen did not generally wear armour, but this increased their vulnerability. Unless pikemen were complemented with firepower, they could not be used to attack cavalry which was on the move. In short, combined arms formations and tactics were necessary, a course that the independence of the mercenary Swiss pikemen made difficult.

> When Malacca fell to Portuguese attack in 1511, the decisive clash was between the sultan's war elephants and a well-coordinated and determined Portuguese force relying on pikes as much as firepower.

SCOTTISH INVASION

The limitations of pikemen had also been indicated at the Battle of Flodden in 1513. James IV of Scotland had invaded England with about 26,000 men, the largest force that had ever marched south. The English under the earl of Surrey advanced with about 20,000 men. James took up a defensive position, but casualties from English cannon fire encouraged part of the Scottish army to advance. After a bitter struggle, the English line held. The Scottish centre was then ordered to advance, but their pikemen were unable to develop momentum in attack, and the more mobile English billmen who used a kind of billhook cum halberd that was about two and a half metres (eight feet) long, defeated them, rather as Spanish swordsmen had been effective against pikemen at Cerignola in Italy in 1503. The Scottish centre was further pressed as other English troops, victorious over the Scottish right, attacked it in the rear; and the Scots were routed, with heavy casualties including James.

Pikes were also used by Europeans expanding across the world. When the major South-East Asian entrepôt of Malacca fell to Portuguese attack in 1511, the decisive clash was between the sultan's war elephants and a well-co-ordinated and determined Portuguese force relying on pikes as much as firepower.

By the second half of the 16th century, the pike was increasingly used in a defensive fashion, with a deployment linked to protecting musketeers, although this did not preclude attack by pikemen. A number of surviving narratives of basic, tactical combat in the Dutch Revolt indicate that pikemen would still frequently charge opposing enemy infantry. A late example of the use of pikes in offensive infantry tactics is provided by the importance of the weapon to the army of Charles XII of Sweden, up until its defeat by Peter the Great of Russia at Poltava in 1709.

Handguns were used in conjunction with pikes until the spread of the socket bayonet around 1700, although the ratio in which they were normally employed during the period changed significantly in favour of guns. The use of the pike-musketeer combination was seen particularly in the European battles of the first half of the 17th century, especially in the Thirty Years War (1618–48) and the English Civil War (1642–6).

From the early 18th century, the pike ceased to be a regular weapon and, instead, became one advocated by those who challenged established conventions. Marshal Saxe, the leading French general of the 1740s, was interested in the idea of reviving the pike, as was the *Encyclopédie* (1751–65), the repository of fashionable French opinion. The use of the pike was seen as a more vigorous means of fighting than reliance on firepower, but the latter continued to prevail.

REVOLUTIONARY PIKEMEN

There was also interest in the pike among revolutionaries. This was seen, for example with the War of American Independence (1775–83). Pikes as weapons first appeared in the general orders for George Washington's army in July 1775, and were last mentioned in August 1776, and they were also used by Pennsylvania Associators from August 1775. That month, Benjamin Franklin drew up a memorandum for the Pennsylvania Committee of Safety, supporting the use of pikes in the rear one or two ranks of units, 'the spirit of our people supplies more men than we can furnish with firearms … each pikeman to have a cutting sword and, where it can be procured, a pistol'. Virginian riflemen sent to help North Carolina in the spring of 1776 were also armed with pikes.

Fortunately for the revolutionaries, this attempt to compensate for the absence of sufficient firearms was not tested in battle: unwieldy pikemen would have presented easy targets for regular musketeers. This was also true for Britain, where in 1779, fears of French invasion and a sense of national emergency led George III to consider distributing pikes to 'the country people'.

The French Revolution led to renewed French interest in the vigour of attacks with cold steel, and in the pike, which was seen as a people's weapon; but this ideological commitment did not extend to any actual adoption of it. The pike was used by Irish insurgents in 1798 when they rose against British rule, but they were rapidly defeated. At the Battle of New Ross, advancing Irish pikemen were finally beaten back by cannon.

ZULU WARFARE

Pikes, however, were not the only thrusting weapon still in use. Heavy, thrusting spears were still of value in some areas, although they were not as long as pikes. They were employed widely in the 18th century, and were also used in the 19th. For example, in southern Africa, the Zulus under Shaka, their chief from 1816 to 1828, used them with considerable effect. Shaka changed Zulu tactics, replacing light throwing *assegais* (javelins) by the *i-klwa*, a heavier, thrusting spear, and emphasizing speedy assault and shock tactics. The success of the Zulu crescent formation was made possible by brave and disciplined troops led by effective officers. Shaka forced defeated peoples to become Zulus – their clans were absorbed, and Zulu aggression and expansion led other peoples to migrate in the *Mfecane* (Time of Troubles). Yet, the Zulus, who were also the last successful major force to use shields, were also to be revealed by the British in 1879 as an anachronism in the face of Western firepower. Despite this ultimate failure, thrusting weapons had had a long history of effective use.

Gunpowder

'Printing, gunpowder, and the mariner's
needle... these three have changed the
whole face and state of things
throughout the world.'

EDWARD GIBBON WAS TO CLAIM THAT GUNPOWDER 'EFFECTED A NEW REVOLUTION in the art of war and the history of mankind', a bold but not overly fantastical claim. Gunpowder weaponry developed first in China. We cannot be sure when it was invented, but the correct formula for the manufacture of gunpowder was possibly discovered in the ninth century, and effective metal-barrelled weapons were produced in the 13th century.

Each of these processes in fact involved many stages. With the gunpowder mixture of sulphur, charcoal and saltpetre, it was important to find a rapidly burning formula with a high propellant force; while with cannon, it was necessary to increase the calibre and to move from pieces made of rolled sheet iron reinforced with iron bands to proper castings.

As with other infantry weapons, gunpowder was used at sea as well as on land. In 1161, gunpowder bombs fired by catapults helped the navy of the Song of southern China to defeat that of the Jurchin Jin.

Knowledge of gunpowder was brought from China to Europe in the 13th century, although the path of diffusion is unclear. In the following century, it was used in cannon. The first 'handgonnes' date to an English reference of the 1380s, and they were known elsewhere in the 1370s. Gunpowder presented the ability to harness chemical energy, and cannon indeed have been referred to as the first workable internal combustion engines.

However, it is also necessary to underline several problems with the idea of a gunpowder revolution. First, although gunpowder provided the basis for different forms of hand-held projectile weaponry and artillery – such as muskets and arquebuses – the technique of massed projectile weaponry was not new, as the English use of longbowmen demonstrated. Therefore it is possible to regard gunpowder weaponry as an agent, not a cause, of changes in warfare, especially so in battle tactics on land.

CHEMICAL AGENTS

It is important to appreciate that gunpowder itself posed serious problems if its full potential as a source of energy was to be used successfully. For a long period, cannon were not strong enough to make proper use of gunpowder. This did not change until the 15th century, with the development,

Early use of gunpowder

Ninth century
Gunpowder is possibly invented by the Chinese during the Tang Dynasty – made from a mixture of sulphur, charcoal and saltpetre. First used as a firework

904
First possible use of gunpowder by the Chinese in warfare as incendiary projectiles called 'Flying Fires'

13th century
The Chinese use mortars fired from bronze tubes.
Knowledge of gunpowder probably reaches Europe via Islamic Spain

1331
Sultan Muhammad IV besieges the Spanish city of Alicante using a machine which threw iron balls with fire

1346
First use of cannon in Europe at the Battle of Crécy

1453
Sultan Mehmed II uses more than 60 large cannon at the siege of Constantinople

around 1420, of a more effective type of gunpowder, which provided the necessary energy but without dangerously high peak pressures. The chemical nature of the gunpowder reaction also caused problems when, for example the chemical reagents lacked consistent purity.

Problems came too from the supply of the ingredients. Charcoal and sulphur were relatively abundant, but about three-quarters of the weight was provided by saltpetre and this was difficult to source. The export of saltpetre by Elizabeth I of England to the Saadian Moors of Morocco in 1578 helped to doom Sebastian I of Portugal's crusade at the Battle of Alcazarquivir. Sebastian was killed, most of the Portuguese bases were captured, and thereafter, no effective Western military pressure was brought to bear upon Morocco until the French did so in the 1840s.

The nature of the projectiles used with early gunpowder weapons was a particular issue. Spherical shot generates a large amount of aerodynamic drag (resistance), essentially because its wake is disproportionate to its cross-sectional area. This ensured that the projectiles of the 15th and 16th centuries lost speed at a high rate: on average about three times faster than modern bullets. Lower speed meant less kinetic energy on impact, and thus less penetrative power. In addition, effects arising from their spin lessened their accuracy as a result of the 'Magnus Effect', in which spinning spheres generate both a low pressure and a high pressure zone. This characteristic could not be counteracted by skill in firing. This was why musket 'balls' were superseded by bullets in the 19th century.

CANNON AND HANDGUNS

Many states, nevertheless, benefited from switching to gunpowder weaponry, which increased markedly in value in the 14th and 15th centuries. For example, the replacement of stone by iron cannon balls, the use of better gunpowder, and improvements in cannon transport , all raised artillery capability. The Ottoman Turks initially relied on mounted archers, but, in the second half of the 14th century, they developed an infantry that became a centrally paid standing army, eventually armed with field cannon and handguns. This helped lead them to a series of victories that ensured that they became the key power in the pivotal area where Europe, Asia and Africa met. In 1473, at the Battle of Baskent, the effect of Ottoman cannon and handguns on Türkmen cavalry led to victory over Uzun Hasan, the head of the Aqquyunlu confederacy (which ruled modern Iran and Iraq), the forces of which centred on cavalry armed with bows, swords and shields.

At Chaldiran (1514), Shah Isma'il I, the Safavid leader, who had overthrown the Aqquyunlu, was in turn defeated by Selim I 'the Grim', the Ottoman sultan. The Safavid army was of the traditional Central Asian nomadic type – horsed archers – but the Ottomans were a more mixed force, and included handgunners and cannon. Although the Safavids had used cannon in siege warfare, they had none at Chaldiran. Cultural factors were important: the Safavids thought firearms cowardly and, initially, adopted cannon with reluctance, preferring to use them for sieges, not battles. Thanks to their firepower and numerical superiority, the respective importance of which is difficult to disentangle, the Ottomans won a crushing victory over the Safavid cavalry. Ironically, the cannon were also important for another reason: chained together, they formed a barrier to cavalry charges. In response to their defeat, the Safavids in 1516 created a small unit of musketeers and gunners. The Portuguese, from their base at Ormuz in the Persian Gulf, were to provide the Safavids with cannon.

Gunpowder weaponry also helped Ottoman expansion in other directions. In 1516,

Ottoman firepower played an important role in the defeat of Mamluk heavy cavalry at the Battle of Marj Dabiq, although other factors also came into play. The Ottoman army was far larger, and the governor of Aleppo, who commanded the Mamluk left flank, made a secret agreement with Selim, and abandoned the battle at its height. This led to the Ottoman capture of Syria, and was followed in 1517 by the Battle of al-Rayda, which led to the conquest of Egypt and the end of the Mamluk empire. From then on the Ottomans dominated Egypt until the close of the 18th century. Initially they left there a garrison of 5,000 cavalry and 500 arquebusiers, which may be an indication of their respective importance.

Egypt served as a base for the projection of Ottoman power down the Red Sea, where Aden was incorporated in the Ottoman system in 1538, and also along the coast of North Africa, although Cyrenaica (eastern Libya) only acknowledged its authority in 1640.

Like the Safavids, the Mamluks had put a premium on cavalry and did not associate the use of firearms with acceptable warrior conduct. Firearms were seen as socially subversive, and hostility to the use of musketeers obliged successive Mamluk sultans in 1498 and 1514 to disband musketeer units they had raised. This helped weaken them when the Ottomans attacked.

There were also major Ottoman gains at the expense of Christian Europe in the late 15th and early 16th centuries, including, in the former, in Albania, Serbia and Greece. At Mohacs in 1526, firepower played a major role in the defeat of the heavy cavalry of Louis II of Hungary by the Ottomans under Selim's successor, Suleiman the Magnificent. The Hungarian cavalry attacked, pushing back the lighter Ottoman cavalry, but was stopped by the infantry and cannon, whose fire caused havoc. The Hungarians, their dynamism spent, were then attacked in front and rear by the more numerous Turkish forces, and routed. Louis II was killed.

OTTOMAN DOMINANCE

This victory was followed by Ottoman dominance of most of Hungary, and in 1529 Suleiman advanced to besiege Vienna, although it did not fall. Gunpowder weaponry had helped transform the geopolitics of the region.

The Ottomans were not alone in their growing use of gunpowder weaponry. The first reference to gunpowder warfare in Muscovy relates to the use of a small cannon in the defence of Moscow in 1382. By 1480, the Muscovites had deployed arquebusiers, and by 1494 Italian cannon-founders recruited by the Tsar had established a cannon-casting yard and powder yard in Moscow. At the Battle of Orsha in 1514, the Poles scored a decisive victory over a larger Muscovite army by using artillery and arquebusiers. This was the first significant Muscovite defeat attributable to superior enemy firepower, and it pushed Vasilii III to begin developing his force of arquebusiers.

In his *Decline and Fall of the Roman Empire* (1776–88), Edward Gibbon wrote 'the military art has been changed by the invention of gunpowder; which enables man to command the two most powerful agents of nature, air and fire … Cannon and fortifications now form an impregnable barrier against the Tatar horse.' Indeed in 1783 the Russians conquered the Crimea, the basis of Tatar power. Firearms had helped transform the relationship between infantry and cavalry, and between the settled states of Europe and the steppe peoples.

Cannon

'... the enemy's battery opened fire upon us and raked us through and through... so accurate was the fire that each discharge of the cannon stretched some of my men on the ground.'

A French officer's account of the Battle of Schellenberg, 2 July 1704 where casualities on both sides amounted to 15,000

MANY BELIEVE THAT IT WAS THE OMINOUS SOUND OF CANNON FIRE THAT HERALDED THE TRUE END OF THE MIDDLE AGES. Cannon certainly had a major impact in sieges and battles. Initially, they were prized for their ability to destroy stone walls and thereby raise sieges. In this role, cannon replaced siege engines, which were now regarded as less effective and more unwieldy. Cannon also came to play a major battlefield role by the mid-15th century in European campaigns.

Outgunned powers were overcome. Cannon helped Charles VII of France defeat the English in the last stages of the Hundred Years War when they were crucial in the rapid capture of English fortresses in 1449–50 and also in the defeat of English armies with longbowmen: at Formigny, the last major clash in Normandy (1450), and in Castillon, the last major clash in Gascony (1453). This new weapon certainly delivered a decisive military verdict.

This was also the case with the conquest of the kingdom of Granada, which culminated with the surrender of the fortress of Granada itself in 1492, the last stage in the Reconquista of Spain and Portugal from the Moors. This victory owed something to divisions within the kingdom, but the Spanish use of largely German-manned artillery was also important. The Moors were outgunned, and the Spaniards successfully developed their gunnery, adopting offensive artillery tactics.

In 1453, the Ottoman Turks, under the skilful direction of Sultan Mehmed II, used gunpowder weaponry in capturing Constantinople (Byzantium, Istanbul). A centre of power that had resisted non-Christian attack for nearly a millennium had finally fallen. The artillery that drove off the Byzantine navy and breached the walls of Constantinople served notice on military attitudes and practices that had seemed to protect 'civilized', i.e. Christian, society. Relying on fortifications to thwart 'barbarians', to offset their numbers, dynamism, mobility and aggression, no longer seemed a credible policy. Siege artillery continued to play a role in the Ottoman advance, as at the successful siege of Modon in 1500 where 22 cannon and two mortars fired a total of 155 to 180 shots per day. The Ottomans also used cannon in gains from non-Christian powers. In 1461, on the southern shore of the Black Sea, the Isfandiyari fortresses of Koyulhisar and Karahisar fell after bombardment.

IVAN THE TERRIBLE

Conversely, Christian powers also used cannon successfully, as in 1552 when Ivan IV, the Terrible, of Muscovy successfully besieged Kazan, the capital of the most northerly Islamic state. The Russian triumph owed much to their first large-scale use of artillery: the Russians had 150 cannon, compared to the 70 in the city. This artillery-based campaign was far more fruitful than those mounted by Ivan in the winters of 1547–8 and 1549–50. In these, the Russians had had to leave their artillery behind because of heavy rains, and ended up campaigning with an exclusively cavalry army that was of no use in investing the fortress of Kazan.

Within Christendom, cannon were used to overthrow fortifications. In 1513 Norham Castle, a strong English position, fell rapidly to the cannon of James IV of Scotland. James II himself had died in 1460 during the siege of English-held Roxburgh when a cannon exploded.

Cannon could also be used by rulers to coerce and overcome recalcitrant cities and aristocrats, as in Scotland in 1456 when the earl of Douglas's castle at Threave surrendered in the face of a 'great bombard'. The gateway of Powis Castle in Wales was 'burst quite in pieces' by cannon in 1644. Batteries of cannon bombarded the Anabaptist-held city of Münster in Germany, which fell in 1535. Castles and walls suddenly seemed redundant or in need of strengthening.

As a result, rulers invested in cannon, and not only in Europe. In the 1530s, Bahadur Shah of Gujarat, in western India, used his state's maritime wealth to fund a large army equipped with new cannon, while, in southern India, Sultan Ibraham Qutb Shah of Golconda financed an artillery corps in the 1560s and 1570s thanks to his monopoly of diamonds. Abbas I of Persia (r. 1587–1629) created a 12,000-strong corps of artillerymen with about 500 cannon.

> Elizabeth I of England ordered her forces besieging Edinburgh in 1560 to take care to collect and bring back the cannon balls they had used.

As ever, the impact of new weapons has to be appreciated in a context that includes an understanding of their limitations. New machines of war have often enjoyed an impact on the imagination greater than that on the battlefield, particularly if their use is accompanied by dramatic sounds and sights. Contemporaries were certainly impressed by the impact of gunpowder weaponry, and a celebrated passage by the Florentine historian Francesco Guicciardini emphasized the impact of the horse-drawn cannon used by Charles VIII of France when he invaded Italy in 1494–5. The French used iron cannon balls. This enabled smaller projectiles to achieve the same destructive impact as larger stone shot, and therefore permitted the employment of smaller, lighter and more manoeuvrable cannon, which were mounted permanently on wheeled carriages. The artillery impressed most contemporary Italians, leading the Aragonese in 1495 to begin to cast iron balls in the Naples arsenal, and the Venetians to order the new cannon (100 wheeled 6- to 12-pounders) the following year.

SIEGE WARFARE

However, detailed study of the campaign indicates that some of the bolder claims on behalf of this enhanced capability have to be questioned. In 1494, the key Tuscan frontier fortresses assailed by the French – Sarzana and Sarzanello – in fact repulsed the attacks, and the French were able to advance only as a consequence of a treaty negotiated by Piero de' Medici. The following January, Montefortino near Valmontone was stormed and sacked, without any apparent use of cannon. A bombardment did make a crucial breach in the walls of Monte San Giovanni, permitting its storming that February, but the bombardment of the Castel Nuovo in Naples was not as effective. Ten days of cannon fire inflicted only limited damage, the French ran short of iron balls and gunpowder, and the surrender of the garrison reflected exhaustion and division, rather than the inexorable pressure of cannon fire. One should also not underestimate the role of mining (as opposed to cannon) in sieges.

Running short of gunpowder and cannon balls was a frequent problem for any protracted siege. Elizabeth I of England ordered her forces besieging Edinburgh in 1560 to take care to collect and bring back the cannon balls they had used. In the French Wars of Religion (1562–98), this problem of supply hit the royal army in 1573 in the unsuccessful siege of the Huguenot (Protestant) La Rochelle, which finally fell, after another lengthy siege, in 1628. In these wars, there were a large number of sieges, and they became more important to the course of the conflict than at the outset. In 1569, the fortress of Lassy fell after the walls had been breached. However, bombardment could only do so much. In 1568,

> Artillery was employed on the battlefield both to silence opposing guns and, more successfully, in order to weaken infantry and cavalry units. Grape and canister shot were particularly deadly; they consisted of a bag or tin with small balls inside which scattered as a result of the charge, causing considerable numbers of casualties at short range.

the Huguenots stormed Chartres, the walls of which had been breached by their nine cannon, but the attack was defeated and the breach sealed.

On the battlefield, the relative immobility of cannon restricted their usefulness, as did their need for clear firing lines: indirect fire was not possible but the close-packed nature of military formations presented the cannon with ready targets. The targeting became more accurate, because of the addition of trunnions on which the barrel could turn in a vertical plane, changing the angle of fire, and the development of moveable carriages.

The battlefield use of artillery increased considerably during the 17th and 18th centuries. By the end of the Seven Years War (1756–63), Frederick the Great of Prussia (r. 1740–86), who had not, initially, favoured the large-scale use of artillery, was employing massed batteries of guns. Cannon became more mobile and standardized: the Austrians in the 1750s and the French, under Jean-Baptiste Gribeauval, from the 1760s, were the leaders in this field. The greater standardization of artillery pieces led to more regular fire, and encouraged the development of artillery tactics away from the largely desultory and random preliminary bombardments of the 17th century, in favour of more efficient exchanges of concentrated and sustained fire.

ACCURACY IMPROVES

Artillery was employed on the battlefield both to silence opposing guns and, more successfully, in order to weaken infantry and cavalry units. Grape and canister shot were particularly deadly; they consisted of a bag or tin with small balls inside which scattered as a result of the charge, causing considerable numbers of casualties at short range.

Mobility was also increased in the 18th century by stronger, larger wheels, shorter barrels and lighter-weight cannon, and more secure, mobile gun carriages. Accuracy was improved by better sights, the issue of gunnery tables, the introduction of inclination markers and more effective casting methods. The rate of fire rose thanks to the introduction of pre-packaged rounds. Horses were harnessed in pairs instead of in tandem.

The theory of war advanced to take note of these changes. In his *De l'usage de l'artillerie nouvelle dans la guerre de campagne* (Paris, 1778), the Chevalier Jean du Teil argued that the artillery should begin battles and should be massed for effect. This certainly influenced Napoleon, who was to use cannon with great effect.

Gunned Warships

'He (Winston Churchill) is still fighting
Blenheim all over again. His only answer to
a difficult situation is "send a gun-boat."'

ANEURIN BEVAN, 1951

CANNON WERE TO MAKE A MAJOR DIFFERENCE IN THE WAGING OF SEA BATTLES. Warships could provide excellent mobile artillery platforms superior to anything on land, and one vessel might carry a heavy firepower capacity comparable to that of an entire army.

Cannon were possibly carried on the Ming Chinese fleet from the 1350s, and a series of seven expeditions was sent into the Indian Ocean between 1405 and 1433. The largest ships carried seven or eight masts, although claims that they were nearly 120 metres (400 feet) in length is questionable, not least as the dimensions do not correspond to the figures for carrying capacity, tonnage and displacement. Nevertheless, they were probably the largest wooden ships built up to then and, thanks to watertight bulkheads and several layers of external planking, they were very seaworthy. The Chinese reached Aden and Mogadishu, and successfully invaded Sri Lanka in 1411 on the third expedition. However, the Chinese gave up this long-range naval power projection in the mid-15th century; and, in its place, that capability was taken by the Western Europeans.

The cannon carried on some European warships from the middle of the 15th century were made to allow for their use at sea in a fashion distinct from that of land-based weaponry: an important aspect of specialization; although one should not exaggerate the differences between guns for siege warfare, fortifications and naval warfare. To a large extent, they were interchangeable, although the mountings had to be changed. Field artillery was different as it had to be as light as possible.

Medieval naval warfare had been dominated by coming alongside and boarding, and this tactic continued to play a role. The rising importance of firepower, however, led to a shift towards stand-off tactics in which the focus was on bombardment.

STAND-OFF TACTICS

Heavy guns were carried in the Baltic, and by English and French warships from the early 1510s. Carvel building (the edge-joining of hull planks over frames), which spread from the Mediterranean to the Atlantic and Baltic from the late 15th century, replaced the clinker system of shipbuilding using overlapping planks. This contributed significantly to the development of hulls, which were stronger and better able to carry heavy guns. Also, their size grew. The English *Henry Grace à Dieu* (also known as the *Great Harry*) had a 1514 specification of 186 guns. The Danish, French, Lübeck, Maltese, Portuguese, Scottish, Spanish and Swedish navies all included ships of comparable size during the course of the century.

The Portuguese were the first systematically to exploit heavy cannon to fight stand-off actions against superior enemies, a development often incorrectly claimed for the English at the time of the Spanish Armada (1588). In northern Europe, the shift towards stand-off tactics can be seen by contrasting the Anglo–French War of 1512–14, in which the fleets fought in the English Channel in the traditional fashion, with the gunnery duel in which they engaged off Portsmouth in 1545. This shift had important implications for naval battle tactics, and it further encouraged the development of warships primarily as artillery platforms.

Firepower was vital to the expansion of European commercial and political power. Large sailing ships with two masts required fewer bases than galleys, which relied on rowers: their crews were smaller, and they could carry more food and water. Able, as a result, to transport a substantial cargo over a long distance at an acceptable cost, these ships, thanks to their cannon, could also defend themselves against attack. The Portuguese were the first to deploy full-rigged sailing ships which were strong enough to carry heavy wrought-iron guns capable of sinking the lightly built vessels of the Indian Ocean, leading to victory over the fleets of Calicut (1502), Japara (1513) and Gujarat (1528).

The Egyptian and Turkish vessels that sailed to the west coast of India, or between the Red Sea and the Persian Gulf, were different in their long-distance capability and were also less heavily gunned. An Egyptian fleet, partly of galleys, sent from Suez in 1507 and supported by Gujarati vessels, initially defeated a greatly outnumbered Portuguese squadron at Chaul in 1508, but, in turn, was largely destroyed by Francisco de Almeida at Diu in 1509. The 72-strong Ottoman fleet sent to Diu in 1538, the largest fleet the Ottomans ever sent to the Indian Ocean, failed to win Gujarati support and was repulsed. In 1554, an Ottoman fleet was heavily defeated by the Portuguese off Muscat.

The heavier armament of the Portuguese warships was crucial in the face of the numerical advantage of their opponents, who had more ships and more men. The Portuguese initially relied on the caravel, a swift and seaworthy, but relatively small, ship, and the *nau* or 'great ship', a very large carrack-type vessel. They subsequently developed the galleon, a longer and narrower ship with a reduced hull width-to-length ratio, which was faster, more manoeuvrable, and capable of carrying a heavier armament. The cannon were fired from the side of the vessel, a change that owed much to the development of the gunport. These cannon could inflict serious damage near the waterline, and hole opposing warships.

INCREASED FIREPOWER

A range of factors was involved in effectiveness and there was a need for trade-offs between mobility, armament and firepower. Balancing cargo capacity, even for warships, sea-keeping quality, speed, armament types and numbers, optimum size of crew and draught, all depended on the role of the ship. European purpose-built warships were typically 500-600 tonnes in the second half of the 16th century, while armed merchant vessels were in the 200-300 tonnes range. Portuguese East Indiamen relied on their vast size for defence, while Spanish *fregates* relied on speed for their safety.

The wooden warship equipped with cannon, whether driven by sails, muscle power (for galleys), or both, was the single most costly, powerful and

The spread of gunned naval warfare

1350s
Ming Chinese fleet carries cannon

1502
Portuguese sailing ships carry heavy wrought-iron guns and their fleets win victory over Calicut (1502), Japara (1513) and Gujarat (1528)

1508
Egyptian galleys carrying guns defeat Portuguese squadron at Chaul

1554
Ottoman fleet defeated by Portuguese superior fire-power off Muscat

1588
Spanish Armada carries cannon on carriages designed for land use. The Spanish are outmanoeuvred by English gunners with compact four-wheeled gun carriages

1592
The Japanese fleet is defeated at Battle of the Yellow Sea by Korean 'turtle' oar-driven boats protected with metal plates and Korean *panokseons* which could carry up to 20 cannon compared to one or two on Japanese ships

technologically advanced weapons system of the period. The introduction of large numbers of cannon on individual ships made maritime technology more complex, and increased the operational and fighting demands on crews. The construction, equipment, manning, supply and maintenance of a fleet needed considerable financial and logistical efforts. The construction of large warships using largely unmechanized processes was an immense task and required huge quantities of wood. The capital investment required was also formidable. Maintenance was also expensive, as wood and canvas rotted and iron corroded. Warships therefore demanded not only, by the standards of the day, technologically advanced yards for their construction, but also permanent institutions to manage them.

The success of European warships as maritime cannon-carriers increased with the development of gunports just above the waterline, and of waterproof covers for them. This ensured that guns could be carried near the waterline, as well as higher up, thus reducing top-heaviness and increasing firepower. Furthermore, developments in gun founding and gunpowder made comparatively light guns more effective and increased the number of cannon that could be carried. As early as 1518, the standard armament of a Portuguese galleon was 35 guns, and when the Japanese ruler Toyotomi Hideyoshi planned his invasion of Korea in 1592, he unsuccessfully attempted to hire two Portuguese galleons. European naval strength transformed the Indian Ocean by making it possible for Europeans to sustain their new presence and to defend new trade routes.

THE SPANISH ARMADA

Within the European world, naval warfare served to record and also advance changes in relative strength between the states. This was particularly so with the rise of English power. The outbreak of war between England and Spain in 1585, ensured that the mighty Philip II of Spain now had an enemy that could only be decisively attacked by sea. He sought to do so in 1588 by sending a major fleet, the Armada, of 130 ships to cover an invasion of England from the Spanish Netherlands (modern Belgium). Sailing through the English Channel, the Spanish warships were harried by long-range English gunnery, although this did little damage. With the advantage of superior sailing qualities and compact four-wheeled gun-carriages, which allowed a high rate of fire, the English fleet suffered even slighter damage, although it was threatened by a shortage of ammunition. Many of the Spanish guns were on cumbersome carriages designed for use on land.

The Spanish fleet reached Calais, but was disrupted there by an English night attack using fireships, and the English fleet then inflicted considerable damage in a running battle off Gravelines. A strong southwesterly wind drove the Armada into the North Sea. The Spanish commanders then ordered a return to Spain via the hazardous north-about route around the British Isles, only for many of the ships to be wrecked as a result of storms. Conflict between gunned warships had thus contributed to a major defeat for Spain.

Four years later, the Japanese fleet was defeated at the Battle of the Yellow Sea by a Korean fleet commanded by Yi Sun-Shin that included some of the more impressive warships of the age: Korean 'turtle ships', oar-driven boats, possibly covered by hexagonal metal plates in order to prevent grappling and boarding, that may also have been equipped with rams. The Japanese rapidly deployed cannon on their warships, and used them with effect in 1593 and 1597, but in 1598 the Koreans were supported by a Chinese fleet under an artillery expert and, that year, the Japanese fleet was defeated at the Battle of the Noryang Straits: the Koreans appear to have had a lead in cannon. This battle helped bring the Japanese invasions of Korea to an end. Gunned warships were undoubtedly key to naval conflict off Asia as well as off Europe.

Matchlock Muskets

'Battles are won by superiority of fire.'

FREDERICK THE GREAT

HANDHELD FIREARMS WERE TO TRANSFORM INFANTRY WEAPONRY replacing archers and javelins. The handgun was developed from the 1380s. By the 1420s and 1430s the powder was ignited by a length of slow-burning match, a lighted fuse. As a consequence, it was possible to separate aiming the handgun from igniting the gunpowder charge, and this improved their value. By 1450, the Burgundians and the French had them in large numbers. They were effective, as in the Burgundian wars against the Liègeois, and in the Swiss and Burgundian wars of 1475–7. By then the term used was 'arquebus'.

Handguns had greater penetrative power than bows, although they had a limited range, a low rate of fire, were affected by bad weather, and were harder to make. Handguns were also heavier than bows, slower to fire, less accurate (and inconsistent in their accuracy), and less reliable. Much of their effectiveness depended on the individual soldier who might be a poor shot – or be unsteady in igniting or otherwise firing. Handguns were also not straightforward to use from horseback as this made it more difficult to load, aim and fire them. Furthermore, they needed a supply of powder and shot, which was not easy to obtain, and unlike arrows, was not recyclable.

Rather than assuming that a shift from bows to arquebuses took place because they were deadly weapons, it might be that it was their lower cost, compared to the crossbow, that led to their adoption. Furthermore, the spread of the weapon appears to have been due not to an instant acceptance of its overwhelming capability, but to its use in a particular niche and by a specific group: militia who guarded city walls, a protected role that compensated for their battlefield vulnerability. From this, the use of the arquebus spread, but it is no accident that the weapon was most effective in the Italian Wars (1494–1559) when employed in concert with field fortifications, as at the battles of Cerignola (1503), Ravenna (1512) and Bicocca (1522). Interlinked problems with range, accuracy and killing power helped guide the tactical use of firearms, as part of a learning curve which entailed an appreciation of their limitations.

ADVANTAGES OF THE ARQUEBUS

The arquebus was adopted only slowly in France and the British Isles, and in the former initially appeared less valuable than a pike-heavy cavalry combination. The French willingness to use cannon, but not arquebuses, is a reminder of the need to be cautious in referring to the impact of gunpowder. The experience of battle did not automatically encourage the use of firepower: the French battlefield success in Italy in 1494–6, campaigns in which they had very little small arms firepower, did not lead them to change the structure of their army.

Nevertheless, the arquebus was increasingly seen as bringing an advantage. At the Battle of Pavia (1525), Spanish arquebusiers, supported by pikemen but fighting in the open, rather than depending on field fortifications, inflicted heavy losses on attacking French cavalry and Swiss pikemen. At the same time, it is not easy to use Pavia as an example of the effectiveness of particular arms because, even more than most battles, it was confused, thanks to the effects of heavy early morning fog, and also many of the advances were both by small units and unco-ordinated; while the surviving sources contain discrepancies. Alongside arguments about the superiority of the arquebus against cavalry, it is necessary to make due allowance for Spanish effectiveness in the small-unit engagements they sought where they responded to French advances by moving out of danger and bringing firepower to bear, in particular to enfilade Francis I and the French cavalry – that is by firing along the line end to end.

At Pavia the terrain was also crucial – part of the battlefield was poor cavalry country, not least in the early-morning fog. Marshy and brush-covered ground delayed the French cavalry, making them targets for Spanish fire. This emphasizes the role of generalship. Francis played to the Marquis of Pescara's advantage in a way that enabled the Spaniards to use their army to maximum advantage. The French advance also masked their artillery. This was the very opposite of cavalry-artillery co-ordination and artillery played little part in the battle.

The musket, a heavier version of the arquebus, capable of firing a heavier shot further and therefore penetrating armour, was first used in Europe from the 1520s and its use became widespread by mid-century. Heavier shot is less affected by deviations than lighter shot, and this made it more accurate although at long range it was still poor. The lead ball of about 15 grams (half an ounce) that was fired could kill at 180 metres (200 yards), but it was very difficult to hit a target at more than 55 metres (60 yards).

THUNDERBOLTS

The arquebus and musket were also important outside Europe. The Incas and Aztecs had neither gunpowder nor firearms, and the Incas referred to the shot fired by arquebuses as *Yllapas*, meaning thunderbolts. The Spanish conquerors of the Inca and Aztec empires benefited from their weapons, although a number of other factors were responsible for their success. When the explorer Magellan arrived at Cebu in the Philippines in 1521, he showed off his firearms and armour in order to convince the population of his potency – this backfired when the bright metal of his armour only proved to make him an easier target and he died in a minor skirmish.

Non-Western forces could also use matchlocks, and their rulers sought to acquire them. In 1472, Uzun Hasan, the ruler of Aqquyunlu Persia, requested arquebuses and cannon from Venice, and by 1478 he had both. The Ottomans provided muskets and musketeers to their allies, including the Uzbeks and the Khan of the Crimea. In 1541, Portugal sent 400 musketeers to the aid of Ethiopia, a Christian kingdom, which was under attack in a *jihad* launched by Ahmad ibn Ibrihim al-Ghazi, the ruler of Adal on the Horn of Africa. With their aid, the

The spread of handheld firearms

1380s
First developed in Europe
1450
The arquebus is in use by the French and Burgundians
1478
Persian forces are supplied with arquebuses by Venice
1494–1559
The Italian Wars saw firearms widely used
1520s
Muskets first seen in Europe
Early 16th century
In South America the Spanish conquistadors use firearms to help crush the Aztecs and Incas
1541
Portugal sends 400 musketeers to aid Ethiopia
1543
Portuguese traders supply the Japanese with guns
Late 16th century
The Mughal emperor, Akbar (r.1556–1605) equips his infantry with muskets

Ethiopians defeated Ahmad who then turned to the Ottoman Turks. They provided him with 900 musketeers and ten cannon, with whose help he defeated the Ethiopian-Portuguese army in 1542. The leading Indian ruler of the 16th century, the Mughal emperor Akbar (r. 1556–1605), equipped his infantry with muskets, and took an interest in their improvement, maintaining a special collection which he tested himself.

The first effective guns in Japan were brought by Portuguese traders in 1543. They were widely copied within a decade, as Japan's metallurgical industry could produce muskets in large numbers, and firearms certainly played an important role in war there from the mid 1550s. However, the most significant changes in Japan occurred earlier, from the late 15th century, as the pace of warfare within Japan accelerated. These changes included larger armies, a greater preponderance of infantry, sophisticated tactics and systems of command; and developments in weaponry, especially the spear and armour. Cannon were used in Japan from at least 1551, although they did not become important until the last quarter of the century. Firepower led to a stress on defensive tactics at the Battle of Shizugatake in 1583, and on the island of Kyūshū in 1587, where Hideyoshi's forces were deployed beyond entrenchments.

Firearms could also play a major role in battles elsewhere. In India, at the First Battle of Panipat in 1526, Babur, the Mughal leader, employed both matchlockmen and field artillery successfully against the cavalry of the Lodis, whose armies did not use firearms. This was followed by the establishment of Mughal power in northern India. At Haldighati in India in 1576, a Mughal force defeated a Rajput army because only the Mughals had musketeers, and they and the Mughal archers killed the elephant drivers who were crucial in their opponent's army. At Tondibi in West Africa in 1591, Moroccan musketry defeated the cavalry of Songhai, leading to the overthrow of its empire.

BALANCE OF POWER

However, the situation was not always as favourable for the use of muskets outside Europe. In general, handgunners were treated badly in South East Asia and the East Indies; they were not members of the warrior élites. For muskets to be effective they had to be used in a controlled manner to provide concentrated fire. The discipline and drill did not match social assumptions about warfare in the region, because they subordinated the skill and social rank of individual soldiers to the collective disciplined unit.

China had lost the lead in handheld firearms, and matchlocks were introduced there, probably both from the Turks, via the Muslims of Xinjiang, and from Portuguese merchant adventurers, either directly or via Japanese pirates. In the 1550s, the Ming rulers of China used large numbers of men armed with traditional weapons – bows, lances and swords – to capture the Chinese bases of the *wako* (Japanese pirates) who attacked the coasts of the Yellow Sea. Nevertheless, despite the large scale use of traditional weapons, the spreading use of firearms pushed up Chinese military expenditure during the century.

In Africa, the use of firearms was restricted by the limited availability of shot and powder and, in the early 17th century, no more than 500 musketeers took part in any one Ethiopian expedition. Although muskets had not yet made an impact in many parts of the world, they had, nevertheless, already caused a major shift in the nature of firepower which greatly helped alter the balance of power both between and within states.

Cavalry Pistols

'Without cavalry,
battles are without result.'

NAPOLEON

FIREARMS TRANSFORMED CAVALRY WARFARE when the invention of the wheel-lock mechanism in the 1520s spread throughout Europe. Unlike the arquebus or musket, which required a lighted fuse, the wheel-lock relied on a trigger-operated spring which brought together a piece of iron pyrites or flint, and a turning steel wheel. The contact produced sparks which ignited the gunpowder in the pistol's pan. This priming charge then forced the main charge through a touch-hole in the barrel. So the glowing (and all too easily extinguished) match of the arquebus or musket could be dispensed with.

The wheel-lock mechanism was more expensive and delicate than the matchlock and more difficult to repair, but it was better suited to the needs of the cavalry as it required only one hand to operate. Similarly, the stabbing weapons of cavalry were lighter than those of the infantry. Wheel-lock mechanisms worked the pistols used by cavalry. And they could fire the pistols while moving and, as three pistols could be carried, fire several shots before reloading. This was useful because reloading in action was not easy. Tests at the time claimed an 85 per cent success rate for hitting a man-sized target at 27 metres (30 yards). In addition, the muzzle velocities of pistols were sufficient to cause nasty wounds if soldiers were hit.

Pistols were used by the Imperial (Austrian) troops in their campaigns in Italy in 1544 and Germany in 1546. They were used in the tactic of the *caracole*, in which ranks of cavalry advanced in order and fired their pistols, before wheeling away from the enemy. This manoeuvre, however, was criticized by commentators, not least because they claimed that pistoleers fired from too far away and that it discouraged cavalry from closing with opposing infantry. It was understandable that they fired from too far, because pistols were outranged by infantry firearms. On the other hand, accuracy was so low that the standard advice was to touch the opponent with the muzzle before firing. The *caracole* must actually have been very effective against formations that were mostly pike, and presumably was devised with that in mind. The fact that *reiters* – the armoured cavalrymen – continued to be hired as mercenaries into the 1580s would suggest that they had some battlefield value, even though military commentators have mostly been very critical of them.

As with other weapons, the effectiveness of cavalry firepower depended on the opposing response. Although infantry could offer more formidable firepower, they could be stationary and unsupported, which left the initiative with the cavalry. In 1552, German cavalry armed with pistols defeated the French at Saint Vincent. These horsemen rarely wore full suits of armour and, although such suits continued to be produced, their functional value declined.

FRENCH WARS OF RELIGION

Cavalry played a major role in the French Wars of Religion (1562–98). At Coutras (1587) and Ivry (1590), Henry of Navarre's cavalry used a mix of pistol and shock tactics. At Ivry, the cavalry charged with their swords after firing their pistols. Henry's cavalry broke through the opposing formation and, once the army of the Catholic League had been defeated, their infantry was crushed with heavy casualties.

Cavalry played a greater role in the Wars of Religion than they had done in the Italian Wars (1494–1559) where their role, though considerable, had been greatly restricted by pikemen, field entrenchments and arquebusiers. In the French Wars, cavalry tended to be well-equipped and experienced, and the role of infantry was less prominent than in the Italian Wars. The cavalry in question carried firearms and were not heavily armoured.

Pistoleers were a challenge to heavy cavalry, not least if the pistoleers did not employ the *caracole*, but, instead, fired and then rode into the opposing ranks using swords. The lance still employed by French heavy cavalry was insufficiently flexible as a weapon for such a struggle. The lance ceased to be used by the Dutch, as did long swords that could be used for thrusting.

Nevertheless, the impact of pistoleers in Eastern Europe was limited. In 16th-century Russia, the cavalry increasingly did not use the bow, but the switch was largely to swords, not pistols. Cavalry played an important role in battle and campaigning in Eastern Europe, both tactically and operationally. The Poles won cavalry victories over the Swedes at Kokenhausen (1601), Reval (1602), Kirchholm (1605), and, over a much larger Russo-Swedish army, at Klushino (1601); although at Klushino, the firepower of the Polish infantry and artillery also played a major role. At Kirchholm and Klushino, the mobility and power of the Polish cavalry, which attacked in waves and relied on shock charges, nullified its opponent's numerical superiority, and the Poles were able to destroy the Swedish cavalry before turning on their infantry. The Poles responded to the Swedish *caracole* by charging in with their swords and smashing the Swedish formation. Exposed once the cavalry had been driven off, the Swedish infantry suffered heavily. At Kirchholm, they lost over 70 per cent of their strength.

This is a powerful reminder of the need to avoid an account of European military development solely in terms of improvements in infantry firepower. Similarly, at Konotop (1659), Russian cavalry was heavily defeated by that of the Crimean Tatars, Cossacks and Ukraine. The Russians lost largely due to poor reconnaissance and generalship: they let their main corps get lured into a swamp.

Polish cavalry tactics influenced those further west, not least thanks to commanders such as Pappenheim who had served in Poland. Gustavus Adolphus of Sweden (r. 1611–32) emulated Polish cavalry tactics, although with less of a commitment to the shock value of an attack at full gallop than was at first claimed by scholars. Gustavus also did not take up the lance.

> In the 1700s, John Churchill, duke of Marlborough made the British cavalry act like a shock force, charging fast, and he used a massed cavalry charge at the climax in his victories over the French at Blenheim (1704), Ramillies (1706) and Malplaquet (1709).

ENGLISH CIVIL WAR

Instead, there was a transition from the *caracole*, with Gustavus at times ordering his first line to fire only one pistol, so that they could rapidly move to the use of their swords, with the lines behind maintaining momentum by only using their swords. At Breitenfeld (1631), a victory that broke the tide of Catholic success in the Thirty Years War (1618–48), the Swedish cavalry on their right flank overcame their Imperialist opponents and drove them from the field, before turning on the Imperialist centre. The cavalry conflicts on the flanks were also crucial at the Battle of Lützen between the Swedes and the Imperialists (1632), and in key battles in the English Civil War (1642–6), such as Marston Moor (1644) and Naseby (1645), in both of which the Parliamentary forces beat the Royalists. Similarly, in the Thirty Years War in the late 1630s, the duke of Saxe-Weimar used his heavily cavalry-based army to fight in an aggressive fashion.

In the 18th century the proportion of cavalry in European armies declined as a result of the heavier emphasis on firearms. Cavalry was principally used on the battlefield to fight cavalry; cavalry advances against unbroken infantry were uncommon, although when they occurred, as with the British victory over the French at Salamanca in Spain in 1812, they could be dramatic. Cavalry also played a major role in securing victory at Blenheim (1704), Hohenfriedberg (1745), Soor (1745), Rossbach (1757), Kunersdorf (1759) and Warburg (1760). In the 1700s, John Churchill, duke of Marlborough made the British cavalry act like a shock force, charging fast, and he used a massed cavalry charge at the climax in his victories over the French at Blenheim (1704), Ramillies (1706) and Malplaquet (1709). Napoleon massed cavalry to help breakthroughs, as with Murat's charge through the Russian centre at Eylau in 1807, while ruthless cavalry follow-ups after victory were also important in consolidating success, as after Napoleon's victory over the Prussians at Jena in 1806.

A DEVASTATING VOLLEY

Nevertheless, the battlefield role of cavalry declined markedly. Cavalry was about three times more expensive than infantry, of limited value in hilly terrain and the enclosed countryside that was becoming more the norm and less effective in the face of infantry armed with flintlock muskets and bayonets. French cavalry charges on the British infantry at Fontenoy (1745) were stopped by musket fire before they could reach the British lines, while at Minden (1759) advancing British infantry drove back French cavalry. Scottish Jacobite Highlanders stopped outnumbered British cavalry at Falkirk in 1746: when the two lines were within ten to 15 yards of one another, the Highlanders fired a devastating volley which disordered the British cavalry. Then the clansmen drew their swords, charged and hacked at the horses' legs.

The social prestige of cavalry service still remained important. This was not simply a matter of the officers. The French cavalry of the 18th century attracted a higher quality of recruit and received better pay.

Cavalry conflict remained important in the transoceanic spread of Western power. One British participant in the defeat of the Mysore forces near Seringapatam in 1792 recorded the attack of the Mysore forces:

> The glittering of the swords in a bright sunshine, and the flashes of the firearms, on both sides, was grand and awful. Our cavalry soon found their overmatch and were obliged to give way in a masterly manner wheeling outwards to the right and left into the rear, by a signal from Colonel Floyd, a moment when the Bengal battalions came up between the two divisions and gave their fire, and perhaps saved the whole corps.

The respective role of firepower and shock, whether on foot or on horse, were to remain a key tactical issue for a long period.

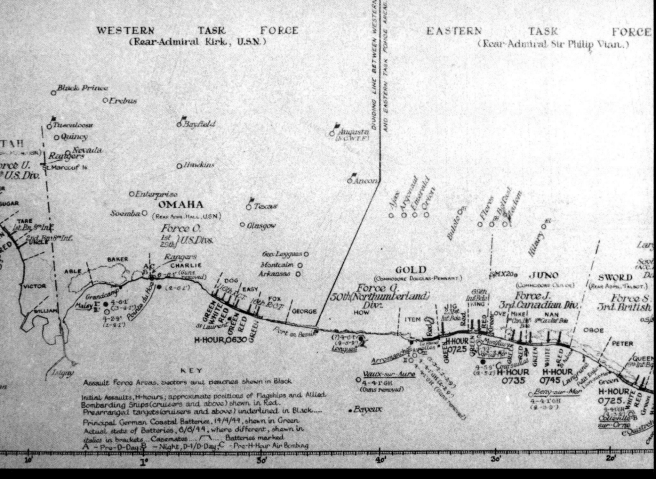

Maps

'The Admiralty sent the *Beagle* to South
America with Darwin on board not because
they were interested in evolution but because
they knew that the first step to understanding
(and, with luck, controlling) the world was to
make a map of it.'

STEVE JONES

THE EXIGENCIES OF WAR DROVE THE DEVELOPMENT OF MAPPING and, in turn, the availability of maps helped greatly in strategic, operational and tactical planning. Much of the history of cartography centres on its military rationale and application, and much of the cartography was prepared under military aegis, or for military purposes. European powers played the key role, but military concerns also led to mapping by non-Europeans. For example, a tradition of cartographic reconnaissance developed in the Ottoman (Turkish) army in the second half of the 15th century, while numerous forts were mapped in South Asia.

Within Europe, the skill base required for cartography was not restricted to the military, but it was the military that had the ability, resources and need to survey and map large areas at various scales. European maritime hegemony from the 16th century rested, in part, on cartographic developments that showed the world's surface on a flat base in a manner that encouraged the planned deployment and movement of forces.

MERCATOR'S PROJECTION

In 1569, the Fleming Gerhardus Kramer, his name Latinized as 'Mercator' (1512–94), produced a projection that treated the world as a cylinder, so that the meridians were parallel rather than converging, as they really do, on the poles. As a result, the poles were expanded to the same circumference as the Equator, greatly magnifying temperate land masses at the expense of tropical ones. Taking into account the curvature of the Earth's surface, Mercator's projection kept angles and thus bearings accurate in every part of the map, so that a straight line of constant bearing could be charted across the plane surface of the map, a goal that was crucial for navigation. However the scale was varied and size was distorted. But this was not a problem for European rulers and merchants keen to explore the possibilities provided by exploration and conquest in the middle latitudes to the west (America) and to the east (South Asia). Europeans had to sail great distances if they were to fulfil the commercial logic of distant possessions and trading opportunities, and to use force to gain more. The map projection they used made most sense in terms of the employment of the compass, and of maritime directions and links, especially in the mid-latitudes.

Cartography was a crucial aspect of the ability to synthesize, disseminate, utilize and reproduce information that was central to European hegemony. Aside from their role in planning, maps served to record and replicate information about areas in which the Europeans had an interest and to organize, indeed centre, this world on themes of European concern and power. The Mercator projection highlighted the imperial world of Portugal and Spain, and was an appropriate prefiguring of the Spanish success under Philip II in creating the first global empire: the first empire on which, as a result of the establishment of Spanish bases in the Philippines from the 1560s, the sun literally never set.

CONTROL OF THE SCOTTISH HIGHLANDS

Mapping was also important to conflict within the West. For example in 1747–55, the British surveyed the mainland of Scotland at the scale of 1:36,000 to produce a map that would, it was hoped, enable the army to respond better to any repetition of the Jacobite rising of 1745. This was the cartographic equivalent to the road and fortress-building policies of the same years. Thus, Scotland, more specifically the Highlands, was to be controlled in a variety of related ways. Fortresses anchored the governmental position at crucial nodes, roads radiating from them offered approaches into the Highlands, and maps provided guidance in the planning and use of force, resulting in the prospect of a strategic, Scotland-wide response to any future Jacobite uprising. None, in the event, occurred.

Large-scale military surveys ensured that reliable detailed maps were produced for much of Europe from the 18th century. For example, the Austrians, who ruled Sicily between 1720 and 1735, used army engineers to prepare the first detailed map of the island. Eighteenth-century French military engineers, such as Pierre Bourcet, tackled the problems of mapping mountains, creating a clear idea of what the Alpine region looked like. This was to help the French when they invaded Italy from 1792.

BRITISH ORDNANCE DEPARTMENT

The lengthy warfare of the French Revolutionary and Napoleonic period (1792–1815) further encouraged the military production and printing of maps. The British Ordnance Department mapped the British Isles in part as protection against the threat of French invasion, while the duke of Wellington used a mobile lithographic printing press in the Peninsular War (1808–14), an example of the military employing new technology, and one that reflected the need for maps.

In the American Civil War (1861–5), field commanders used maps extensively, although, at the outset, they were affected by a shortage of adequate ones. Commercial cartography could not serve military purposes, and thus the armies turned to creating their own map supplies. By 1864, the United States Coast Survey and the army's Corps of Engineers were providing about 43,000 printed maps annually for the Union's army. In that year, the Coast Survey produced a uniform, ten-mile-to-the-inch-based map of most of the Confederacy east of the Mississippi river.

Technology served the cause of the combatants, lithographic presses producing multiple copies rapidly. The production of standard copies was crucial, given the scale of operations, especially the need to co-ordinate forces over considerable distances. This was true in the American Civil War not only of campaigns, but also of battles. The scale of the latter was such that it was no longer sufficient to rely completely on the

Maps

1569
Projection map produced by the Flemish Garhardus Kramer (1512–94) known as 'Mercator'. It highlights the imperial world of Spain and Portugal

1720–35
Austrians prepare the first detailed map of Sicily which was then under their control

1747–55
The British survey the mainland of Scotland with maps at the scale of 1:36,000 to help them control any future rebellions

1808–14
The duke of Wellington uses a mobile lithographic printing press during the Peninsular War to aid his campaign

1861–5
In the American Civil War, the Union army produce about 43,000 maps annually

1914–18
During World War I, the British Expeditionary Force was responsible for more than 35 million map sheets

1939–45
British Ordnance Survey produced over 300 million maps and American Army Map Service over 500 million maps for the Allied war effort

field of vision of an individual commander and his ability to send instructions during the course of the engagement. Instead, in a military world in which planning, and, more specifically, staff specifically appointed for planning, came to play a greater role, maps became key to this process.

The use of maps for tactical reasons was amply displayed in World War I (1914–18). Accurate large-scale maps were crucial for trench warfare, not least for artillery locating its targets. This reflected the growing emphasis on indirect fire. There was a massive growth in military mapmaking which, in part, depended on new technology, with an extensive use of aerial photography: cameras mounted on balloons and aeroplanes.

WARTIME PRODUCTION

There was also a great expansion in the production of maps during the war. When the British Expeditionary Force (BEF) was sent to France in 1914, one officer and one clerk were responsible for mapping, and the maps were unreliable. By 1918, the survey organization of the BEF had risen to about 5,000 men and had been responsible for more than 35 million map sheets. No fewer than 400,000 impressions were produced in just ten days in August 1918.

This experience ensured that in World War II (1939–45), mapmakers were rapidly recruited for the war effort by the combatants. For example, Armin Lobeck, Professor of Geology at Columbia University, produced maps and diagrams in preparation for Operation Torch, the American invasion of French North Africa in 1942. Quantity was also important. For example, the British Ordnance Survey produced about 300 million maps for the Allied war effort although its offices in Southampton were badly bombed by the Germans in 1940. The American Army Map Service produced over 500 million copies.

The role of air power, on land and at sea, ensured that many maps had to become more complex. The role of air power dramatized for Americans by the Japanese attack on Pearl Harbor in 1941, led to a new sense of space, which reflected both their vulnerability and the awareness of a new geopolitical relationship. The Mercator Projection was unhelpful in the depiction of air routes: great circle routes and distances were presented poorly in this projection, as distances towards the Poles were exaggerated.

More specifically, air power led to an enhanced demand for accurate maps in order to plan and execute bombing and ground-support missions. Maps of target areas were eagerly sought. Existing printed maps were acquired and supplemented by the products of photo-reconnaissance and other surveillance activity. The Germans used British Ordnance Survey maps as the basis for maps including information from photo-reconnaissance that they produced to guide their bombers during World War II. Such photo-reconnaissance was also important for the creation of maps for amphibious and land operations, as with the German attack on the Soviet Union in 1941, which was preceded by long-range reconnaissance missions by high-flying Dornier Do-215 B25 and Heinkel He-111s, or the Allied invasion of Normandy in 1944.

After World War II, photo-reconnaissance mapping continued as an aspect of the Cold War, but it was superseded by satellite-based photography. The creation of digital maps of the world's surface was important for the use of cruise missiles. As always mapping technology continues to interact with the development of weaponry.

Trace Italienne

'The art of defending fortified places
consists in putting off the moment of
their reduction.'

FREDERICK THE GREAT

FORTIFICATIONS WERE TRANSFORMED THANKS TO THE CHALLENGE OF CANNON and were designed both to lessen the cannon's impact as a siege tool and to provide platforms for the use of cannon by the defence. This helped give the latter a major advantage and was particularly important to the spread of Western power, enabling relatively small groups of men to consolidate their gains.

The new fortification technique, known as the *trace italienne*, developed from the late 15th century and was a response to the improvements in artillery that century. Fortifications were redesigned to provide lower, denser and more complex targets. In place of the high stone walls that were vulnerable to cannon, came lower, thick walls, strengthened with earth, that tended to absorb much of the shot. The walls sloped downwards in order to deflect cannon fire. Large masonry buttresses were constructed behind the walls to help them withstand bombardment.

Along all the walls at regular intervals were bastions, generally quadrilateral, which served as artillery platforms. Angled to the walls, they provided effective flanking fire, thus overcoming the problem for medieval circular towers of dead ground that could not be covered. The bastions could also provide mutual support.

The use of earthworks preceded the full *trace italienne*. They were a relatively inexpensive way to strengthen the defences, not least by making it harder to storm them, while they also absorbed artillery fire. Free-standing interior earth ramparts helped protect Pisa in 1500 and Padua in 1509 from storming. The polygonal bastion, a more expensive undertaking, developed from the 1450s, with important work by Francesco di Giorgio Martini (1439–1501) and the Sangallo family.

BASTIONS AND RAVELINS

New-style fortifications were particularly vital in Italy, being built, for example, at Civitavecchia in 1515, Florence in 1534, Ancona in 1536 and Genoa in 1536–8. The techniques of bastioned works were then spread across Europe by Italian architects and engineers such as those hired by Francis I of France in 1543. In addition to the bastions, the fortifications were strengthened by ravelins, triangular earthen fortifications that supported bastions and the curtain walls.

The Venetians were very quick to use the new military architecture in their overseas *empire da mar*: on Crete, a fort was built at Candia (Iraklion) in the second quarter of the century and a major fortress at Rethymnon after 1573. In the late 16th century, the Austrians improved the leading fortresses in the section of Hungary they had retained, using the cutting-edge expertise of the period: Pietro Ferrabosco, Carlo Theti and other Italian military engineers provided the plans and directed the works. Fortresses at Eger, Ersekujvar, Kanizsa, Karlovac, Komáron and Gÿor (Raab) were built or rebuilt. On another of Christendom's frontiers, the fortress of St Elmo, built in 1552 by a Spanish engineer, brought the defences of Malta enough time in 1565 to ensure that the Turkish siege of Valetta ultimately failed.

The Ottomans had no equivalent to the *trace italienne* nor to the extensive fortifications built by the Spaniards along the coasts of Naples and Sicily, but they did not require any such development as they were not then under attack.

The new fortresses were defended by cannon and states were willing and able to deploy large numbers of them. Newhaven, the English star-shaped fort built outside Le Havre in 1562, was equipped with 19 cannon, 15 brass culverins, 29 brass demi-culverins and 2 cast-iron demi-cannon, plus another 70 smaller, wall-mounted pieces. The fortress fell to the French in 1563 because of the severe impact of plague and because winds in the English Channel delayed relief, not because its walls were breached or stormed.

Fortresses could resist sieges, as Ostend did in 1601–4, before surrendering, but, like castles before them, were also dependent on the general flow of a campaign. In particular, operations in the field (either battles or the decision not to engage) helped determine the fate of a fort, as the garrison could lose the hope of relief.

DUTCH REVOLT

Formal fortifications were not always the key to defensive positions. In the Dutch Revolt against rule by Philip II of Spain in the late 16th century, Dutch defensive successes were not due to the bastion system, but to improvised earthen barricades behind the breaches. The Spaniards found enormous difficulty in carrying out effective siege operations against Haarlem, Alkmaar and Leiden, as the Dutch also had problems at Middleburg due to the extensive nature of the water defences. These made it difficult to build up the forts or sconces for the siege artillery. Mining was also out of the question because of the low water table, which also allowed waterborne re-supply. The Dutch focused on earth defences, rather than the more expensive stone and this ensured that they could be built more rapidly. The earth ramparts were fronted by an outer earthwork called a *fausse-braye* which ensured that the defenders controlled the ditch, while ravelins were supplemented by demi-lunes (half-moon redoubts) and hornworks.

Fortifications played a key role in campaigning in the 16th and 17th centuries. Thus, the War of the Mantuan Succession (1628-31) revolved around sieges, not battles. There was a contested succession for the Duchies of Mantua and Monferrato, that began with a joint Savoyard and Spanish invasion in 1628 when Spain besieged Casale, the capital as well as a major fortress in Monferrato and a crucial point on the western approaches of the Milanese. The Austrians, in turn, sent in troops and, in 1629, they besieged Mantua, which surrendered the following year. French success in relieving Casale, in 1630, led to peace negotiations.

VAUBAN

Investment in fortifications was seen as the key way to consolidate positions. This was particularly the case with France under Louis XIV (r. 1643–1715). Under his predecessor, Louis XIII, there had been major works, for example at Pinerolo, but nothing that compared with the systematic attempt to defend vulnerable frontier regions with new fortifications that his son supported. Appointed Commissioner General of Fortifications in 1678, Sébastien Le Prestre de Vauban supervised the construction of 33 new fortresses, such as those at Arras, Ath, Blaye, Lille, Mont-Dauphin, Mont-Louis and New Breisach, and the renovation of many more, such as Belfort,

'fortresses... subject an enemy to the necessity of attacking them, before he can penetrate further; they afford a safe admission to your own troops on all occasions; they contain magazines, and form a secure receptacle, in the winter time, for artillery, ammunition etc.'

Marshal Saxe 1732

Besançon, Landau, Montmédy, Strasbourg and Tournai. In 1703, Vauban became the first engineer to reach the prestigious rank of Marshal of France.

In essence, Vauban's skilful use of the bastion and of enfilading fire represented a continuation of already familiar techniques, particularly layering in depth, and he placed the main burden of the defence on the artillery, but it was the crucial ability of the French state to fund such a massive programme that was novel. For example, New Breisach, built to control an important Rhine crossing, cost nearly three million livres to construct between 1698 and 1705.

French territorial expansion was directly linked with the construction of fortresses. They were designed to stabilize the frontier to the advantage of France, consolidating acquisitions, and yet also to facilitate opportunities for fresh gains by increasing France's presence in contested areas and safeguarding bases for operations, and the crucial accumulation of stores. In 1732, Marshal Saxe wrote of 'the usefulness of fortresses; they cover a country; they subject an enemy to the necessity of attacking them, before he can penetrate further; they afford a safe admission to your own troops on all occasions; they contain magazines, and form a secure receptacle, in the winter time, for artillery, ammunition etc.'

A MASTER OF SIEGECRAFT

Vauban himself argued that the greater number of fortresses placed an increased premium on siegecraft,

> one can say that in it alone today is the means of conquest and defence, because the gain of a battle only brings temporary acquisitions unless the fortresses are seized… a war waged by sieges exposes a state least and gives the most chance of conquests, and today it is most practised in warfare in the Low Countries, Spain, and Italy, whereas in Germany battles play a greater role because the country is opener and there are fewer fortifications.

He was a master of siegecraft and favoured a systematic approach in which saps were dug in zig-zag patterns to support parallel trenches from which heavy bombardment of the fort could be launched.

Fortified positions were also important in conflict outside Europe. Thus, control of Kandahar, in what is now Afghanistan, was important in the conflict between the Persians and the Mughals of India. Lost by the Mughals to Shah Abbas I of Persia in 1622, Kandahar was regained in 1638, when the Persian commander surrendered, fearing execution by his sovereign. Helped by Mughal weakness in the aftermath of their unsuccessful campaign in northern Afghanistan in 1647, Shah Abbas II recaptured Kandahar in 1648, and Mughal attempts to regain it in 1649, 1652 and 1653 all failed. It was difficult to campaign effectively so far from the centre of Mughal power, and success had to be achieved before the harsh winter set in.

In India itself, the Mughals were more successful. In 1687, their main field army besieged Golconda with its four-mile-long outer wall. Two mines were driven under the walls, but they exploded prematurely. The walls finally fell by betrayal: Mughal forces entered through an opened gateway. The Mughals also captured a whole series of Maratha forts in 1689 and 1700–7. Improvements in fortification continued to ensure that fortresses played a central role in campaigns.

Ships of the Line

'At this period the enemy were forming their
double line in the shape of a crescent.
It was a beautiful sight when the line was
completed: their broadsides turned towards us
showing their iron teeth, and now and then
trying the range of a shot to ascertain the
distance, that they might, the moment we
came within point-blank (about 600 yards)
open their fire...'

MIDSHIPMAN BADCOCK ON THE MORNING OF THE BATTLE OF TRAFALGAR

THE TOWERING AND PICTURESQUE SAILING SHIPS OF THE LINE WERE THE DOMINANT EUROPEAN FIGHTING SHIPS in the 17th, 18th and early 19th centuries. They reflected the development of line-ahead deployment and tactics for warships where they sailed forward together in a line, which encouraged the maximization of broadside power. The stress on ships of the line was a product of the emphasis on fleet actions fought with heavy guns and of the development of specialized warships and, in particular, of the focus on gunfire.

Notable British sailing ships

1509–10
Mary Rose is built at Portsmouth
1512–14
Henri Grâce à Dieu, nicknamed 'Great Harry' built at Woolwich Dockyard
1545
Mary Rose is accidentally sunk during the Battle of the Solent against the French
1577
HMS *Revenge* is launched at Deptford. In 1588 it became Francis Drake's flagship
1637
HMS *Sovereign of the Seas* is launched at Woolwich. She was extravagantly decorated and the £65,000 she cost helped to precipitate the financial crisis which led to the Civil War
1759–65
HMS *Victory* is built at Chatham dockyard and becomes Nelson's flagship at the Battle of Trafalgar. Now preserved in Portsmouth
1784
HMS *Indefatigable* is launched. It inspired C.S. Forester to set his fictional hero Horatio Hornblower on her decks

In part this came about through the development of cheap large-calibre cast-iron guns. The manufacture of such weapons was initially beyond the technological capability of the period, but, from the mid-15th century, firepower was increased by the development of large cannon cast from lighter, more durable and more workable 'brass' (actually bronze, which is an alloy of copper and tin). These cannon were thick enough to withstand the high pressure from large powder charges and were able to fire iron shot with a high muzzle velocity and great penetrative force. As a result, the stone shot used in early cannon was phased out. It had been rendered redundant by the new iron shot, although stone continued to be used where iron was not available.

From the 1540s, cast-iron cannon were produced in England, which was a major source of iron. The Dutch were able to produce cast-iron cannon by the 1600s, but such cannon were preferred for merchantmen not for warships, as they could burst when overheated through rapid firing. Successful cast-iron guns began to be mass-produced in Sweden in the 1610s with the help of Dutch technicians. From the mid-17th century, cast-iron cannon became the leading naval artillery.

BROADSIDE POWER

Conflict at sea increasingly emphasized firepower exchanges. In the Battle of the Downs (1639), a major Dutch victory over Spain in the English Channel, the Dutch kept their distance, preventing the Spaniards from closing and employing boarding tactics. In the ensuing artillery exchange, the Dutch, who employed line-ahead tactics, inflicted greater damage. In 1653, during the First Anglo-Dutch War (1652–4), the English warships were ordered, in their fighting instructions, to provide mutual support. While this was not an order to fight in line ahead, it encouraged the line formation that maximized the broadside power of the entire fleet. The stress on cohesion reflected a move away from battle as a series of struggles between individual ships toward more planned combat, though the nature of conflict at sea, not least the unwieldy nature of the line, made it difficult to maintain cohesion once ships became closely engaged.

More generally, European warships became more heavily gunned. Instead of relying on converted merchantmen, purpose-built warships that were heavily gunned and, accordingly, had strong hulls, were used. This led to a professionalization of naval officership, senior ratings and infrastructure, and also ensured that less heavily gunned vessels were rendered obsolete for fleet actions. In the First Anglo-Dutch War, the English warships were larger than the Dutch and had a higher ratio of cannon per tons. They had many warships that could carry full batteries of 32- or 18-pounders, while the Dutch had only a few that could carry batteries of 18- and 24-pounders. Partly as a result, the English warships dominated the naval engagements.

However, lacking, by modern standards, deep keels, sailing vessels suffered from limited seaworthiness, while the operational problems of working sailing ships for combat were very different from those that steam-powered vessels were to encounter. The optimal conditions for sailing ships were to come from windward in a force 4–6 wind across a sea that was relatively flat; it was more difficult to range guns in a swell. Limitations on manoeuvrability ensured that ships were deployed in line, in order to maximize their firepower, and skill in handling ships in line, or in battle, entailed balancing the wind between the sails of the three masts in order to achieve control over position and speed.

Line tactics and fighting instructions were designed to encourage an organizational cohesion that permitted more effective firepower, mutual support and flexibility in the uncertainty of battle. Tactical practice, however, conformed to theory even less at sea than on land, due partly to the impact of weather and wind. Although experience, standardization and design improvements enhanced performance, there were still significant limitations.

COMMERCIAL RAIDERS

Significant naval forces with a range greater than war canoes were deployed by only a handful of non-European powers, principally the Ottoman (Turkish) empire, the Barbary states of North Africa (Algiers, Morocco, Tripoli and Tunis), and the Arabs of Oman. The ships of these powers approximated to European warships, but lacked their destructive power. The Barbary and Omani ships were commercial raiders that emphasized speed and manoeuvrability, whereas the heavier, slower ships of the line of European navies were designed for battle and battering power. As far as the European fleets were concerned, they required different kinds of ships for line of battle exchanges than for commercial raiding and protection. The emphasis in the latter was on speed, and thus on frigates that were too lightly gunned for the line of battle.

The programmes of naval construction designed to provide this shipping indicated not only the resources of European governments, but also the capability of their military-industrial complexes. Fleets were powerful and sophisticated military systems, sustained by mighty industrial and logistical resources, based in dockyards that were among the largest industrial plants, employers of labour and groups of buildings in the world, for example

Portsmouth, Plymouth, Brest, Toulon, Ferrol, Cadiz and Karlscrona. The 1,095-foot-long ropery opened at Portsmouth in 1776 may well have been the largest building in the world at the time. In 1704, Peter the Great founded the Admiralty shipyard at St Petersburg, and a naval academy followed in 1715. His move of the capital to St Petersburg owed much to its role as a port. As soon as the Russians had seized a Black Sea coastline and the Crimea in 1783, they began to develop bases there, particularly at Kherson, Sevastopol and Odessa.

The size of warships altered in response to requirements. During the 18th century, improvements in seaworthiness, stemming in part from the abandonment of earlier top-heavy and clumsy designs, increased the capability of warships, both to take part in all-weather blockades and to operate across the oceans. The emphasis on maximizing firepower in the late 17th century led to a development of three-decker capability. In the early 18th century, the focus, instead, was on stability, range and versatility, which led to a move toward two-deckers, but three-deckers became important again for the bruising confrontations of the late 18th century. Whereas in 1720, there were only two warships displacing more than 3,000 tons, by 1815 nearly a fifth of the naval strength above 500 tons was in this category. In 1800–15, ships of 2,500–3,000 tons achieved greater importance, whereas those of 2,000–2,500 and 1,500–2,000 tons declined in number.

HEAVIER GUNS

These bigger ships were able to carry heavier guns. Whereas the average ship of the line in 1720 had 60 guns and was armed with 12- and 24-pounders, that of 1815 had 74 guns with 32- and 36-pounders on the lower deck. The invention of a system of ship construction using diagonal bracing in order to strengthen hulls and to prevent the arching of keels, designed by Robert Seppings, was to permit the building of longer two-deckers armed with 80 or 90 guns. The first ship built entirely on this principle, HMS *Howe*, was not launched until 1815. More mundanely, but also as part of a general process of improvement, there were also developments in fittings, for example new patterns of anchors and the first chain cable, as well as iron water-tanks in place of wooden casks.

The British navy was to prove particularly successful in the French Revolutionary and Napoleonic Wars (1793–1815). The fundamentals of British strength were an unmatched commitment to naval power, and a skill in employing it to maximize other advantages and forces. British naval hegemony rested on a sophisticated and well-financed administrative structure, as well as a large fleet drawing on the manpower resources of a substantial mercantile marine (although there were never enough sailors), and an ability to win engagements that reflected widely diffused qualities of seamanship and gunnery, a skilled and determined corps of captains, and able leadership.

British success in naval conflict which culminated at Trafalgar on 21 October 1805 when an outnumbered British fleet wrecked their Franco-Spanish opponents who lost 19 ships of the line, helped transform the Western world. It ensured that the British became the dominant imperial power. Naval strength permitted amphibious operations. These included the capture of the Dutch bases of Cape Town (1806) and Batavia (now Djarkarta, 1811), and of the French bases of Martinique (1809), Réunion (1810) and Mauritius (1810). During the war of 1812–15 with America, British naval strength permitted the launching of amphibious operations and a blockade that hit the American economy. Despite their limitations, ships of the line worked well as the cutting edge of a sophisticated system of naval power.

Bayonets

'The onset of Bayonets in the hands of the
Valiant is irresistible.'

MAJOR-GENERAL JOHN BURGOYNE, 1777

THE FEARSOME EDGE OF BAYONETS WHICH WERE DEVELOPED towards the close of the 17th century, altered warfare in Europe by transforming the footsoldiers' capability in combat. The early plug bayonet, introduced in the early 1640s, was inserted in the musket barrel and therefore prevented firing. This bayonet was based on a weapon used by hunters and was named after Bayonne in southwest France. They were daggers that, if necessary, could be inserted into muskets, making them a useful weapon against boars.

It has been claimed that the French army was using the bayonet by 1642. Use rapidly spread, and by the 1670s specialized units such as dragoons and fusiliers were being issued with bayonets. At the siege of Valenciennes in 1677 the first French bayonet attack occurred. By the 1680s, they were far more common. These bayonets were essentially double-edged dagger blades that were about 30 cms (12 inches) long attached to a handle that was about 30 cms (12 inches) long. This was designed to be the same diameter as the musket's bore. The handle was fixed in position by working the handle into the musket. In 1672, bayonets were issued to a unit in the English army, Prince Rupert's Dragoons.

The plug bayonet was replaced by ring and socket bayonets, developed in the 1680s. These allowed firing with the blade in place. The bayonet was turned and locked in place, providing firmness in combat.

This led to the phasing out of the pike, which was now redundant. Bayonets were a better complement to firearms in fulfilling the pike's defensive role against attacking infantry and cavalry, and also had an offensive capability against infantry and, on occasion, cavalry. Firepower was greatly enhanced as a result of the replacement of pikemen.

A PROTOTYPE BAYONET

This rapid change was largely carried out in the 1690s and early 1700s. The Württemberg troops sent to campaign against the Turks in Greece in 1687–8 had no pikes, and the Saxon and Württemberg armies converted to bayonets between the late 1680s and the mid-1690s. In 1687, the Marquis de Louvois, the French army minister, instructed Vauban to make a prototype bayonet. Given Vauban's skills in fortress design, this was a testimony to his general ability.

Brandenburg-Prussia adopted the bayonet in 1689 and Denmark in 1690. At the Battle of Fleurus (1690), in the Nine Years War or War of the League of Augsburg (1688–97), some German units attracted attention by repulsing French cavalry attacks despite being armed only with muskets and unsupported by pikes. Russia adopted the bayonet from the 1700s. In contrast, the Turks were slow to do so: they were used, and then only in relatively small numbers, from the 1730s.

It had been very complicated to co-ordinate pikemen and musketeers to ensure the necessary balance of defensive protection and firepower. The new system, in contrast, led to longer and thinner linear formations, and shoulder-to-shoulder drill used to maximize firepower, that characterized European infantry in the 18th century, both within Europe and overseas. The use of the bayonet and of flintlocks encouraged the development of offensive tactics, not least because more effective infantry weaponry led European forces to phase out body armour which increased the mobility of their troops. Initially, however, bayonet drills were based on pike drills, with an emphasis on receiving advances. It was not until the 1750s that a new bayonet drill made it easier to mount attacks.

Despite the bayonets, hand-to-hand fighting on the 18th-century battlefield was relatively uncommon and most casualties were caused by shot. At the Battle of Malplaquet in 1709, the bloodiest battle of the War of the Spanish Succession, about two per cent of the wounds suffered by French troops were the result of bayonets. Nevertheless, bayonets were seen as of importance and were added to new weapons. In 1786, Sir William Fawcett, the British Adjutant General, returned to George III two guns the king had sent him, 'the bayonet of that which is intended for the use of the Light Infantry having been made to fix, agreeably to your Majesty's directions'.

BATTLE OF MINDEN

A bayonet charge, preceded by a volley, had become a standard British tactic from the late 1750s – guaranteed to strike fear in the hearts of the enemy. At the Battle of Minden in 1759, a key engagement in the Seven Years War, the courage and fire discipline of the British infantry won the battle. They misunderstood orders, advanced across an open plain, and then repulsed two charges by French cavalry. Most of the cavalry casualties were caused by musket fire, but those who reached the British lines were bayoneted. These charges were followed by a French infantry advance, and then by another cavalry attack, which was again stopped by musketry and bayonets. This victory greatly lessened the possibility of French pressure on Britain's ally, Frederick the Great of Prussia.

> The bayonet was essentially a psychological weapon in most Napoleonic engagements. Firepower caused more casualties and was therefore crucial to the battle's outcome while the bayonet charge exploited any advantage.

The bayonet charge was used with effect in the War of American Independence (1775–83). In 1778, George Wayne of the Pennsylvania Line asked the State Board of War to exchange all the rifles in his division for muskets, 'I don't like rifles. I would almost as soon face an enemy with a good musket and bayonet without ammunition – as with ammunition without a bayonet.' Wayne claimed that riflemen often fled in panic when attacked by soldiers armed with bayonets. The British fear of American marksmen was therefore matched by American dread of British bayonet charges.

THE REDOUBT AT YORKTOWN

As a consequence, the Inspector General, Baron von Steuben, introduced bayonet practice into the drill of the (American) Continental Army at Valley Forge in early 1778, and at the Battle of Monmouth Court House that year the Continental Army made its own bayonet attack. The bayonet was also used by the Americans when storming Stony Point, Paulus Hook, and, in 1781, the Redoubt at Yorktown. The last was a key episode in driving in the blockaded British position at Yorktown, and helped encourage Cornwallis, the British commander to see the position as lost.

With his fine grasp of timing and eye for terrain, Arthur, 1st duke of Wellington subsequently brought the British tactical system to a high pitch of effectiveness. In the Peninsular War (1808–14) in Spain and Portugal, the British launched well-timed bayonet charges as the key element in counterattacks, taking advantage of the disorganization of the French by their approach march and by British fire. Medical records on casualties, and other sources, suggest that the bayonet was essentially a psychological weapon in most Napoleonic engagements. Firepower caused more casualties and was therefore crucial to the battle's outcome while the bayonet charge exploited any advantage.

Bayonet charges also proved crucial to British victories in India, such as Patna (1764) in which bayonets and grapeshot blocked the Indian cavalry, or Wellington's victory over their major enemy, the Marathas, at Assaye in 1803. In this battle, Highland infantry advanced against Maratha cannon. Wellington's dispatch, printed in the *London Gazette* of 31 March 1804, noted the role of the bayonet,

> *the troops advanced under a very hot fire from cannon... they advanced in the best order... against a body of infantry far superior in number, who appeared determined to contend with them to the last, and who were driven from their guns only by the bayonet.*

INDIAN VICTORIES

At Assaye, Wellington, with 4,500 troops, 17 cannon and 5,000 unreliable Indian cavalry, successfully confronted a Maratha army of 30,000 cavalry, 10,000 infantry and over 100 cannon. The Maratha cannon moved fast, were well laid and served, disabled the British cannon and inflicted heavy casualties. Repeated attacks were necessary to force the Marathas to retreat. Casualties accounted for over a quarter of the British force. Crucially, the fearsome Maratha cannon were captured. In a subsequent British victory that year at Argaon, infantry advances at bayonet point against effective Maratha cannon were again crucial.

Battles, however, were not the sole factor. The Marathas were weakened by a poor command structure and by lack of money and the absence of regular pay destroyed their discipline and control. The British were stronger under both counts.

The role of the bayonet continued to be important in European successes. In the First Sikh War (1845–6) in northern India, British victories at Mudki (1845), Firozshah (1845) and Sobraon (1846) owed much to infantry actions with the bayonet. Similarly, bayonets were used in Africa, as with the British attack and victory over the Egyptians at Tel-el-Kebir in 1882, the key success in the conquest of Egypt.

In the 20th century, the long sword bayonets of the 19th century, which had helped to affect the balance of bayonets, were replaced by shorter bayonets. These were designed to provide soldiers with a weapon in the event of hand-to-hand conflict, which was uncommon but, nevertheless, still sometimes occurred, as in the Falklands War of 1982.

Despite the accentuation of firepower in the 20th century, not least with the application of machine guns and rapid-firing artillery, bayonets remained a feature of many (although by no means all) hand-held firearms.

Flintlocks

'Now the tug of war began. As they could only get away a few at a time, not only were the bayonets used but many were the fractured skulls by the butts of firelocks.'

The Letters of Private Wheeler, August 1813

WHEN MATCHLOCKS WERE REPLACED BY FLINTLOCKS THERE WAS A MAJOR SHIFT IN FIGHTING MECHANISMS that increased the speed and reliability of musket fire – as well as leading to more casualities – with major consequences for effectiveness both in Europe and further afield. Flintlocks made musket fire less dependent on the weather. Matchlocks had been affected by wind, rain and general humidity, the last proving a particular problem in the tropics.

The flintlock musket, in which power was ignited by a spark produced through striking flint against steel, was more expensive, but lighter, not requiring a rest, less unreliable, easier to fire and more rapid-firing than the matchlock. The rate of fire, helped by the spread of pre-packaged paper cartridges, which provided the correct amount of powder, almost doubled. Without the hazard of the burning matches previously used to ignite powder, musketeers were also able to stand closer together which increased the firepower per length of unit frontage.

The first form of flintlock ignition system was the snaphaunce lock from the mid-16th century. The more classic flintlock was developed from this in the early 17th century. These flintlocks were used for hunting and for pistols, but it took time to introduce the flintlock as a major military weapon because of the cost.

The Austrians adopted the flintlock in about 1689, the Swedes from 1696, the Dutch and English by 1700. In England, all the new regiments raised from 1689 were equipped with flintlocks. The new Land Pattern Musket could be fired at least twice a minute and weighed 450 grams (one pound) less than the matchlock previously used. Like the earlier adoption of the arquebus, the spread of the flintlock was not instantaneous, however; unsurprisingly so, as the cost of one was equal to the annual wages of an agricultural labourer. Nevertheless, as an indication of competitive pressure and the clear advantage of the flintlock, its adoption was quicker than that of the arquebus. Although French regulations permitted the use of flintlocks by some soldiers from 1670, matchlocks were not completely phased out until 1704. This delay probably owed much to cost, but the Nine Years War (1688–97) led the French to decide to change over, and this was decreed by an ordinance of December 1699. Many Swedish units continued however to use older forms of firearms.

HIGH CASUALTY RATES

The impact of the flintlock was magnified by the replacement of the pike by the bayonet. The bayonet-flintlock musket combination altered battlefield tactics, helping to lessen the role of cavalry, and ensured that casualty rates could be extremely high, particularly as a result of the exchange of fire at close quarters between lines of closely packed troops, the formation chosen to maximize firepower. The Battle of Malplaquet in 1709 was particularly bloody.

Low muzzle velocity led to dreadful wounds, because the more slowly a projectile travels the more damage it does as it bounces off bones and internal organs. Soldiers fired by volley, rather than employing individually aimed shot. It was difficult in any case to aim in the noise and smoke of a battlefield, while, as a further problem, the heavy weight of muskets led to musket droop: firing short. There was also bruising as a result of the recoil.

Even in perfect conditions, effective range for an individually aimed shot was only about 45 metres (50 yards), although this has been the subject of much debate, and some certainly claimed to have shot accurately at greater distance. It was unusual to exceed three shots a minute. Accuracy was compromised by the nature of the barrel (unrifled and, in order to avoid fouling by powder and recoil, generally a loose fit for the shot, and therefore of limited accuracy); the shot, often elliptical and thus unlikely to travel as designed; the slow lock time; the need for rapid firing; and the absence of accurate sights. The shot, as a result of the significant windage, 'bounded' down the bore and might leave the barrel in any direction, a process known as balloting. Worn flints and blocked touch-holes caused misfiring, while reloading became more difficult as the bore fouled.

The flintlock musket was the foremost weapon for both civilian and military use for over 200 years
- It first became generally available in the 1630s
- Terms applying to their use such as 'lock,stock and barrel', 'going off at half-cock' and 'flash in the pan' are still prevalent today
- The British Ordnance Office issued 56,000 muskets between 1701 and 1704
- In 1777 the French supplied 23,000 muskets to the American revolutionaries

The development of iron, instead of wooden, ramrods was believed to increase the rate of musket fire. However, these ramrods often bent and jammed in the musket, or broke or went rusty, and frequent use of the ramrod distorted the barrel into an oval shape.

There were also production problems with the musket. The calibre of individual Prussian muskets ranged between 18 and 20.4 millimetres (three-quarters of an inch to just over an inch), while their length varied by up to 8 centimetres (3 inches). The French 1754-model flintlock was largely handmade, the parts were not interchangeable, and the lock was intricate and difficult to standardize. The situation was still poor in the early 1790s, with very limited interchangeability. In 1757, it was estimated that many French muskets could not fire six times without danger of breaking.

ROLLING FIRE

There were variations in the order in which ranks (lines) or platoons fired, with the Dutch system of platoon fire becoming common during the War of the Spanish Succession (1701–14), except for the French. The Prussians then used a system with each platoon firing separately, producing a rolling fire, although the French continued to fire by ranks until the mid-18th century.

The French commander Marshal Saxe, the victor over the British at the Battle of Fontenoy in 1745, noted that,

> *the present method of firing by word of command, as it detains the soldier in a constrained position, prevents his levelling with any exactness… according to the present method of loading, the soldiers, in the tumult and hurry of an engagement, very seldom ram down their charge, and are also very apt to put the cartridges into the barrel without biting off the caps, by neglecting to do which, many of the arms are of course rendered useless.*

Nevertheless, despite their limitations, flintlocks offered large-scale battlefield firepower, and, as a result, many were made. Between 1701 and mid-1704, the British Ordnance Office issued 56,000 muskets. The main Russian state arsenal at Tula produced an annual average of nearly 14,000 muskets between 1737 and 1778 while, in the 1760s, the French produced

23,000 muskets annually at Charleville and Saint-Etienne. In 1777, France supplied 23,000 muskets to the American revolutionaries. In 1814, the British had 743,000 serviceable muskets in store. European armies were not to experience a comparable change in weaponry until the introduction of rifled guns in the 19th century.

The use of flintlocks was not restricted to European powers, although in many areas, such as Africa and South East Asia, it was spread by their trade. For example, the flintlock replaced the matchlock as the principal European firearm export to the Gold and Slave Coasts of West Africa from about 1690. Local blacksmiths could make copies of these flintlocks. On the island of Madagascar, the modern European firearms acquired through exchange for slaves by the kingdom of Ambohimanga helped in the early 19th century, to enable it to conquer its opponents, whose forces were armed with spears and old guns.

Similarly, in New Zealand, as a result of trade with Westerners, the use of muskets spread from the 1810s. Their high value was such that in 1820 one musket was worth 200 baskets of potatoes or 15 pigs. Maori raiders armed with muskets became increasingly active in New Zealand from 1820. The Nga Puhi from the northern tip of the North Island raided to the southernmost tip of the island, using their muskets to win victories – for example at Nauihaina in 1821 and Totara in 1822.

BATTLEFIELD ADVANTAGE

Flintlocks were also used in conflict among Asian powers. At the Third Battle of Panipat near Delhi in 1761, the Afghan victors over the Marathas included mounted musketeers armed with flintlocks, who had largely replaced mounted archers. Flintlocks, however, were not manufactured in China or South East Asia. The rate of adoption of flintlock muskets was lower and slower in Africa and Asia than in Europe. This ensured that European-armed units had a battlefield advantage over Asian forces.

Flintlocks could also be used with cannon. They were applied to cannon in the British navy from 1778, as a result of the initiative of Captain Sir Charles Douglas, who fitted out his ship, HMS *Duke*, with them at his own expense, leading to faster, more reliable and better-controlled fire. The British benefited at the Battle of the Saints in 1782, and other navies followed suit. This led to an improvement in fighting capability despite the essentially static form that typified 18th-century European navies.

These static forms, however, were challenged by tactical innovation toward the close of the century. At sea, this involved 'breaking the line', which was developed by the British, and on land a similar tactic of column attacks on traditional close-order linear formations. The latter was developed by the armies of revolutionary France and successfully used by them from 1792.

Flintlocks were superseded in the 19th century by ignition systems that were more rapid and more reliable. Nevertheless, they had earlier played a major role in military capability.

Rifles

'My shoulder was as black as coal, from the recoil of my musket; for this day I had fired 107 rounds of ball-cartridge.'

FROM THE JOURNAL OF A BRITISH SOLDIER IN THE PENINSULAR WAR

RIFLES OFFERED GREATER ACCURACY THAN MUSKETS, and they therefore advanced the effectiveness of infantry firepower. Rifled barrels first became important as a battlefield weapon in the late 18th century, particularly in the American War of Independence (1775–83), although they had been used from the 16th century by huntsmen and had indeed been issued to some soldiers.

In 1679, Louis XIV of France ordered that two good shots in each cavalry company should carry rifled carbines, and a large number of rifles was supplied to the French army in the 1680s. Rifles were more accurate than the common flintlock musket, had greater range, and were more appropriate for individually armed fire. The rifling or arrangement of grooves in the barrel led to an aerodynamic spin that aided consistency in flight, and therefore aiming. This was a marked improvement on unrifled barrels.

Riflemen were used in conflict in Germany during the Seven Years War (1756–63). The duke of Richmond recalled a unit of Brunswicker riflemen,

> *800 of this sort of force, being posted by General Imhoff… in a thick wood near Cassel, directly in the front of the French army. It was astonishing the execution they did, and the difficulty with which they were dislodged was inconceivable. They placed themselves two and two behind the trees, and were such admirable marksmen, that as soon as any of the enemy ventured forward, they dropped them. After trying for a considerable time to dislodge these riflemen, the French general was obliged to march up a large body of his infantry with the utmost rapidity, and by that means, with very great loss on his side, at length dispossessed them of the wood.*

In North America, the Kentucky or Pennsylvania rifle was a formidable weapon at long range. Employed as a frontier gun, for hunting and Indian fighting, it was used, however, only by frontiersmen. In 1775, ten companies of riflemen from the Maryland, Pennsylvania and Virginia frontier were raised by order of Congress to help the New England force outside British-occupied Boston. John Adams described them as 'an excellent species of light infantry. They used… a rifle – it has circular… grooves within the barrel and carries a ball with great exactness to great distances. They are the most accurate marksmen in the world.' The riflemen from western Maryland under Michael Cresap were also armed with tomahawks and dressed in hunting shirts and moccasins. Fast-moving, the Marylanders marched 550 miles to Boston in three weeks, then started to thin the British sentries, exacerbating the hemmed-in and depressed atmosphere in the besieged British army. James Murray, a member of the latter, reported,

> *the reasons why so many officers fell is that there are amongst the provincial troops a number of enterprising marksmen, who shoot with rifle guns, and I have been assured many of them at 150 yards, will hit a card nine times out of ten … though these people in fair action in open field would signify nothing, yet over breast works, or where they can have the advantage of a tree (or a rock) and that they may have every 20 yards in this country, the destruction they make of officers is dreadful.*

At Bemis Heights in 1777, the American riflemen under Daniel Morgan concentrated on picking off British officers, Burgoyne, the British commander noting,

The enemy had with their army great numbers of marksmen, armed with rifle-barrel pieces: these, during an engagement, hovered upon the flanks in small detachments, and were very expert in securing themselves, and in shifting their ground. In this action, many placed themselves in high trees in the rear of their own line, and there was seldom a minute's interval of smoke in any part of our line without officers being taken off by a single shot.

The British defeat at Bemis Heights ended General Burgoyne's attempt to cut the rebelling colonies in half by marching south from Canada via Lake Champlain and then along the axis of the Hudson valley toward British forces based in New York. Had this advance succeeded, then the articulation of American power would have been dramatically reduced, and the Americans, instead, would have become more a series of local forces, which the British could have tried to fight, and/or negotiate with separately.

VICTORIOUS REVOLUTIONARIES

At King's Mountain on 7 October 1780, the victorious revolutionaries fought with rifles 'in their favourite manner... an irregular but destructive fire from behind trees and other cover... flying whenever there was danger of being charged by the bayonet, and returning again as soon as the British detachment had faced about to repel another of their parties'. The defeat of pro-British loyalists under Patrick Ferguson in this battle led to a key weakening of the British position in the Carolinas, not least of the left flank of British forces advancing from the south.

Rifles, however, also posed serious problems. A rifle could carry no bayonet, took one minute to load, because the shot had to be pushed down the barrel, and also needed an expert to fire it, of which there were few. The slow rate of fire of the riflemen was not much of a problem if they were sniping, but was a serious difficulty in close-order fighting, if an enemy closed in during a firefight, offsetting his poorer accuracy with sheer volume of fire. Rifles could also be fouled by repeated firing, as indeed could muskets, but the problem was increased in the rifle because the windage (gap between the barrel and the projectile) was so much less. Due to fouling, rifles needed scouring. Rifles were also considerably more expensive than flintlocks partly because they were difficult to produce and required special ammunition.

> Rifles, however, also posed serious problems. A rifle could carry no bayonet, took one minute to load, because the shot had to be pushed down the barrel, and also needed an expert to fire it, of which there were few. The slow rate of fire of the riflemen was not much of a problem if they were sniping, but was a serious difficulty in close-order fighting.

As a result of these characteristics, the rifle was seen essentially as a weapon for special units trained to fight in open order, rather than for close-order fighting in massed formations: one of the crucial features of infantry conflict was that weapons were too heavy and unsophisticated for soldiers either to carry several firearms or to have one multi-purpose weapon. Rifles could be used as smoothbore muskets in order to increase the rate of fire, but the rifled grooves rapidly became clogged. Washington employed riflemen as skirmishers and snipers, not as regular soldiers.

THE BATTLE OF BRANDYWINE

Riflemen could make an important contribution when they had good cover, either natural, for example north of Fort Washington in 1776 or at Bemis Heights in 1777, or artificial, but these situations arose less frequently than is commonly supposed in American folklore with its romantic notions about frontier riflemen and their rifles. The Hessian *Jägers* who served with the British provided them with effective riflemen. *Jägers* were at the front of Cornwallis's column at the Battle of Brandywine in 1777, and their sniping harmed the defenders of Charleston in 1780.

The rifle did not play a major role in military preparedness after the War of American Independence. Although rifles were used in the French Revolutionary and Napoleonic Wars (1792–1815), they were not crucial to them. The British indeed neglected the use of rifled weapons in the period between the American War of Independence and the creation, in 1800, of the Experimental Rifle Corps, which eventually developed into the 95th Regiment.

This was part of the expansion of light infantry, a cause associated in particular with Sir John Moore, the tragic victor of Corunna, who in 1803 was appointed commander of a new brigade at Shorncliffe Camp in Kent which was designed to serve as the basis of a permanent light infantry force. Particular emphasis was placed upon marksmanship. Moore's force was to become the Light Brigade and, subsequently, the Light Division. The British, however, were still unable to use riflemen as effectively as the Americans, and this affected them at the Battle of New Orleans in 1815, with an attack on a narrow front providing a good target for destructive American defensive fire.

A high rate of fire was also a characteristic of another radical innovation, the Austrian *Repetier windbüchse* (repeating air rifle), invented in 1780 and used until 1800. Employing compressed air from a flask in its stock, this air rifle was very accurate, had a good range and was fitted with a 20-shot magazine that permitted a high rate of fire. The gun, however, needed frequent maintenance work, depended on a cumbersome portable air pump for reloading and was not readily mass-produced. In about 1770, Edward Bate of London constructed an air gun in which the air was pressurized via a pump in its butt. These are examples of innovations that were practical but could not be adopted as major battlefield weapons. They reflected the process of experimentation with firearms that was a product of the quest for enhanced effectiveness.

Semaphore and Telegraph

'What hath god wrought?'

NUMBERS 23:23

SAMUEL MORSE'S FIRST TELEGRAPH MESSAGE FROM WASHINGTON TO BALTIMORE,
24 MAY 1844

COMMUNICATIONS HAVE ALWAYS BEEN A KEY PROBLEM FOR MILITARY OPERATIONS. PRIOR TO MODERN SYSTEMS, they were a particular complication for manoeuvres at a distance beyond the range of visual signalling or sending messages by horseback. In the War of American Independence (1775–83), the nature of communications prevented the exercise of close control and made it hard to respond to developments adequately. This seriously accentuated difficulties in co-operation between the British army and navy.

Indeed, many British operations proved an object lesson in the troubles created by poor communications. This was true, for example, of the British campaign of 1777 in North America. Instead of a coherent plan, there were two totally unco-ordinated campaigns. Operating in conjunction with each other, Generals Burgoyne and Howe might have been able to wreck the two American armies, to gain total control of the Hudson valley, to separate New England from the middle and southern states, and to initiate a state-by-state pacification of New England. Operating, however, as two roving columns, they achieved far less even had they been undefeated. In the event, Burgoyne was blocked by the Americans at Bemis Heights. The British forces further south were not informed sufficiently rapidly to be able to provide support.

Nevertheless, there was progress in communications in the late 18th century. Signalling at sea, which was crucial to tactical effectiveness, communications and co-ordinated action, improved from the 1780s. A quick and flexible numerical system of signals was developed by the British. It was generally possible for a lookout to see only about 15 miles from the top of the main mast in fine weather. However, fleets used a series of frigates stationed just over the horizon, and they signalled using their sails, which were much bigger than flags, and, because the masts were so tall, could be seen at some distance over the horizon. This relay system was particularly important for British fleets blockading hostile ports: there would be an inshore squadron of highly manoeuvrable ships which signalled, using a relay of frigates, to the main fleet, which was located a few miles off in greater safety: close inshore operations exposed ships to the danger of running aground.

APPLICATION OF SCIENCE

The semaphore was an important example of the increased application of scientific method to aspects of military affairs during the period of the French Revolutionary and Napoleonic Wars (1792–1815). Claude Chappe developed the semaphore telegraph. He was given official approval in 1793 and the network of semaphore stations, with an average distance of seven miles between each, created from 1794 by the French Revolutionary government was extended by Napoleon to reach Venice, Amsterdam and Mainz, each of which the French controlled. A set of arms pivoting on a post provided opportunities for multiple positions, and these constituted the message. The system had a capacity of 196 different combinations of signs and an average speed of three signs a minute. Furthermore, code could be employed. In favourable weather, one sign could be sent the 150 miles from Paris to Lille in five minutes, although fog, poor weather and darkness gravely limited the system by affecting visibility. Telescopes on the towers were used to read messages and these towers were constructed about five to ten miles apart. No better system of communications was devised until the electric telegraph.

The Paris to Lille route was important because Lille was a key base for French operations on the northeastern frontier. This frontier was under attack from allied forces in 1792–4, and, in turn, the French made decisive advances there in 1794.

The semaphore system was copied abroad, an aspect of the way in which what appeared to be best practice was rapidly diffused within the Western world. Chappe's work was translated into English by Lieutenant-Colonel John Macdonald, a military engineer who did much work himself on improving telegraphy, publishing *A New System of Telegraphy* in 1817. In the 1790s, the British Admiralty in London was linked to the major base of Portsmouth by semaphore, while Sweden also built some stations. Furthermore, nine stations were constructed along the lines of Torres Vedras, fortified positions built in 1810 by the duke of Wellington to protect Lisbon against French attack. In good weather, a message could be sent the 22 miles from the Atlantic to the River Tagus in seven minutes.

Nevertheless, semaphore networks were very limited, in part because of visibility but also because the stations could be attacked. Indeed, British naval forces in the Mediterranean put ashore parties in order to attack the French stations. Napoleon investigated the possibility of a mobile semaphore system for his 1812 invasion of Russia, but it was considered unviable. Most orders and reports were still handwritten and communicated by mounted messengers, both on the battlefield and at the operational level.

MESSAGE TRANSMISSION

The electric telegraph was to prove a far more effective system for the transmission of messages, both overland and across the sea. By 1900, more than 170,000 miles of ocean cables were in use. The electrical telegraph was developed in about 1837, with the electromagnet used to transmit and receive electric signals. The American Samuel Morse also developed the simple operator key and refined their signal code, which became Morse code.

The telegraph offered both rapidity and range. The latter was of great value in the co-ordination of far-flung resources, while, more generally, the telegraph facilitated the practice of strategy. During the Crimean War (1853–6), the European telegraph network was extended to the Crimea, allowing Napoleon III of France to intervene in allied operations, to the understandable irritation of his generals, and William Russell of *The Times* to send home critical reports. The British used the telegraph effectively in countering the Indian Mutiny (1857–8). From 1859, the Prussians sent orders for mobilization by telegraph, and the telegraph was linked to the development of the Prussian rail system. This was very important to the promotion of Prussian

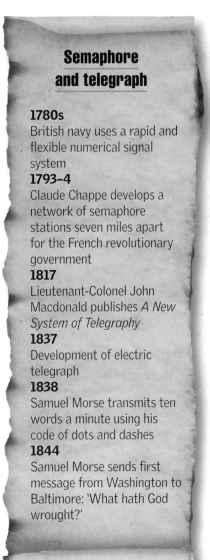

Semaphore and telegraph

1780s
British navy uses a rapid and flexible numerical signal system
1793–4
Claude Chappe develops a network of semaphore stations seven miles apart for the French revolutionary government
1817
Lieutenant-Colonel John Macdonald publishes *A New System of Telegraphy*
1837
Development of electric telegraph
1838
Samuel Morse transmits ten words a minute using his code of dots and dashes
1844
Samuel Morse sends first message from Washington to Baltimore: 'What hath God wrought?'

On the North-West Frontier of India, in the Waziristan rebellion that began in 1937, the tribesmen cut telegraph lines. Meanwhile, field telephones had become important from the late 1890s, being used by the Americans in the Spanish-American War (1898) and by the British in the Boer War (1899-1902). It was, however, the radio that was to make earlier systems obsolete.

mobilization in successive conflicts. Furthermore, during the Austro–Prussian War of 1866 and the Franco–Prussian War of 1870-1, the telegraph was used to help co-ordinate Prussian advances. Earlier, in the American Civil War (1861-5), both sides used the telegraph, which ensured that cavalry raiders sought to cut telegraph lines.

The telegraph was also used to issue instructions for far-flung troop movements. This was very much the case for the British empire, with, for example, the dispatch of troops from India to Abyssinia (Ethiopia) in 1868 and from Ceylon (Sri Lanka) to Natal during the Zulu War in 1879. The former helped in the concentration of forces, and the latter in the movement of reinforcements in what became very much a crisis as a result of over-optimistic British generalship and unexpected Zulu resistance. In 1898, Commodore George Dewey's squadron of American warships was ordered from Hong Kong to Manila by telegraph and its arrival there and victory over the heavily outgunned Spanish squadron at Cavite was to be crucial to the defeat of Spain in the Philippines.

That year, the British won their confrontation with France over competing interests in the southern Sudan, the Fashoda crisis, without conflict, in part by manipulating information about it thanks to their control of the telegraph links; although the major display of British naval strength in European waters was also crucial.

As a result of the value of telegraph systems, their planning became a key aspect of strategic policy. For example, in developing communication routes between Britain and India through the Middle East, the British took great care to run their telegraph route under the Persian Gulf, where it would be difficult to intercept, rather than overland along the shores, where it would be far easier for hostile locals to cut the telegraph. Nevertheless, there were still major problems with running telegraphs under the sea, and it took time to devise the appropriate technology, as the successive attempts to lay cables under the Atlantic indicated.

Telegraph lines were also seen as of an operational importance in colonial operations. As a result, telegraph systems were established to help strengthen the imperial presence, as when the Italians occupied Eritrea in 1885. Due to this, telegraph lines were frequently attacked by insurgents. On the North-West Frontier of India, in the Waziristan rebellion that began in 1937, the tribesmen cut telegraph lines. Meanwhile, field telephones had become important from the late 1890s, being used by the Americans in the Spanish–American War (1898) and by the British in the Boer War (1899–1902). It was, however, the radio that was to make earlier systems obsolete.

A discussion of the semaphore and the telegraph serves to introduce the theme of the impact on war of rapid developments in communications. These were to be of key importance at every level of war, tactical, operational and strategic. Communication systems proved key weapons, as well as important indicators of the modernization of war, and of major changes in how it was waged.

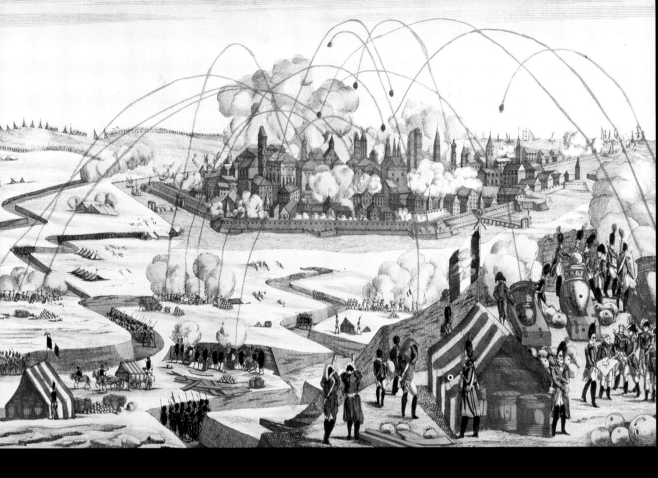

Rockets,
Percussion Caps
and Breech-Loaders

'And the rockets' red glare,
the bombs bursting in air,
Gave proof through the night
that our flag was still there;
O say, does that star-spangled banner yet wave
O'er the land of the free
and the home of the brave?'

FRANCIS SCOTT KEY 'THE STAR-SPANGLED BANNER'
(WITH REFERENCE TO THE BRITISH ATTACK ON BALTIMORE IN 1814)

EXCITING INNOVATIONS IN THE USE OF GUNPOWDER WERE INCREASINGLY SEEN IN EUROPE FROM THE LATE 18TH CENTURY. It took a variety of forms. The rocket represented an attempt to develop an Asian innovation, while the percussion cap was a European one. The British mathematician and gunnery expert Benjamin Robins read papers to the Royal Society on rockets in 1749 and 1750. These were developed by William Congreve (1772–1828), who in 1791 was attached to the Royal Laboratory at Woolwich and spent his own money on his research.

The inspiration was from the Indian use of war rockets by Mysore forces against the British at Seringapatam in 1799. Congreve argued that 'the rocket is, in fact, nothing more than a mode of using the projectile force of gunpowder by continuation instead of by impulse; it is obtaining the impetus of the cartridge without the cylinder, it is ammunition without ordnance, and its force is exerted without re-action or recoil upon the fulcrum from whence it originates'.

A planned British attack by boat-mounted rockets on the French invasion-port of Boulogne was thwarted by adverse winds in 1805, but by then Congreve's rockets had a range of 2,000 yards, and 500 could be discharged at one time from ten launchers. By 1806, Congreve was making larger rockets, which were used in attacks on Boulogne (1806) and Copenhagen (1807). Congreve rockets caused extensive fires and panic when the French-occupied Dutch port of Flushing was attacked by a British amphibious force in 1809. Congreve was then making 32-pounder rockets with a range of 3,000 yards, considerably greater than that of field artillery; but, in 1810, the more mobile 12-pounder rocket was produced.

TERRIFYING EFFECT

Wellington was sceptical because of the difficulty of controlling the rockets' flight, but Commodore Sir Home Popham, who operated on the Biscayan coast in 1812 in the Peninsular War, pressed for them, 'they are admirable... the Spaniards are quite astonished'. Congreve was permitted to raise two rocket companies, one of which served at the Battle of Leipzig (1813). It impressed Alexander I of Russia, but caused fear rather than damage. The rockets had a similar terrifying effect at the British crossing of the River Adour in 1814, but they were seen as decisive in causing the French to retreat. In 1827, Congreve published *A Treatise on the General Principles, Powers, and Facility of Application of the Congreve rocket system, as compared with artillery*. Congreve's rockets were, nevertheless, expensive to produce, as well as inaccurate. Both the French and the Saxons also used rockets.

Rockets did not realize their full potential until the 20th century when a guidance system could be added. This changed the rocket from a glorified firework to a predictable weapon. There were also important developments in rocket power. In the Soviet Union, Konstantin Tsiolkovsky (1857–1936) advanced a theory of rocket flight which encouraged the use of liquid propellants for them, while other work led to the development of the Katyusha multiple rocket launcher. The Germans used long-range rockets in 1944–5, replacing their earlier air attacks, and now without the loss of pilots; but were not able to aim the rockets accurately. V-1s were followed by V-2s, which travelled at up to 3,000 mph and, at the speed and height they used, could not be destroyed by anti-aircraft fire. The V-2 could also be fired from a considerable distance. Large numbers of rockets were aimed at

London, causing much loss of life and fear, although other targets, especially the port of Antwerp once it had been captured by the Allies in 1944, were also hit. The Germans were planning the development of a missile able to hit New York.

PERCUSSION CAPS

In 1807, Alexander Forsyth (1768–1843), a Scottish cleric, patented the use of fulminates of mercury in place of gunpowder as a primer for firearms. Mercury fulminates ignited when struck: there was no need for external fire and thus detonation. The resulting use of the percussion cap, coated with fulminates of mercury, produced a reliable, all-weather ignition system. However, a lack of government support delayed the use and development of Forsyth's invention, and its impact was minor until after 1815. As a result of the use of the percussion cap, it was possible to produce a priming (initial ignition) charge and a main charge that occurred nearly simultaneously, thus greatly speeding up the process of fire.

> In 1807, Alexander Forsyth (1768-1843), a Scottish cleric, patented the use of fulminates of mercury in place of gunpowder as a primer for firearms. Mercury fulminates ignited when struck: there was no need for external fire and thus detonation. The resulting use of the percussion cap, coated with fulminates of mercury, produced a reliable, all-weather ignition system.

Mass-produced metal percussion caps dated from 1822 when Joshua Shaw developed one. Positioned over the fire hole, it ignited the main charge, replacing the flintlock mechanism. This was one of a number of percussion systems which were initially developed for hunters and other civilians, and are a prime example of the extent to which military developments benefited from earlier civilian uses.

Percussion muskets were introduced into European armies, as for example the Austrian army, from 1836. The dramatic reduction of misfires resulted in a great increase in firepower. In the 1840s, percussion-lock rifles followed. Their rifled barrel gave bullets a spin, which led to a more stable and thus reliable trajectory, and, therefore, far greater accuracy and effectiveness than the balls fired by smoothbore muskets. They were also less expensive to manufacture than earlier rifles and, unlike them, could be fitted with bayonets, an important increase in the capability of the rifle as opposed to the musket.

BREECH-LOADERS

Percussion firearms were to be replaced by breech-loaders. Breech-loading firearms had a long genesis. They were used for cannon as early as the fourteenth century. There were also attempts to develop handheld firearms, which became more frequent in the 18th century, mainly for hunters. As breech-loaders did not need to have the shot rammed down the barrel, they could be fired seven times a minute but proved unsuited to frequent use as the loading mechanism was susceptible to clogging by powder. In 1758, the French began to manufacture a breech-loader, but the weapon was a failure and the inventor, Bordier, committed suicide. In the 1770s, Captain Patrick Ferguson produced an improved breech-loader for the British, but this initiative was not developed after his death in 1780. In 1819, a breech-loading rifle, conceived by John

Hall, that employed a breech-chamber, and not cartridges, was adopted by the American army.

Aside from important cost issues, the great constructional problem with the breech-loading rifle was the escape of gas at the breech, and this was the cause of the major delay in its adoption in the 19th century. The potential of the bolt-action breech-loading, Prussian Dreyse *Zündnadelgewehr* or 'needle' rifle (named after its needle-shaped firing pin), a rifle with an hitherto unprecedented high rate of fire, was not fully realized until the 1860s. Although the Dreyse rifle was adopted by the Prussians and first used, with great effect, by them against German revolutionaries in Baden and Hesse in 1849, Prussia was not a combatant in the 1850s. Furthermore, early versions of the Dreyse rifle suffered from design faults, including a brittle firing pin, a bolt action that was liable to jam, and a weak gas seal around the breech. These discouraged borrowing the breech-loading technology compared to the rapid diffusion of percussion-lock rifles.

DEADLY EFFECT

Nevertheless, the Dreyse rifle could be loaded lying down and fired four to seven times a minute, although in practice, it was fired largely from the hip because of the problem of escaping gases. By the 1860s, the accuracy rate was 65 per cent at 300 feet and 43 per cent at 700 feet, although the volume of fire, not its accuracy was seen as crucial.

The Prussians used these rifles to deadly effect against the Danes in 1864 and Austria in 1866. The Danes were still reliant on muzzle-loaders, which were slower and had a shorter range than the Prussian 'needle' rifle. The merits of the weapons on both sides were clearly shown at encounters where Danish losses were far higher, such as the Battle of Lundby. Similarly, in 1866, Prussian infantry fire devastated successive Austrian advances, for example the attacks at Burkersdorf, Rudersdorf and Skalice. The Austrians suffered because they had drawn the wrong conclusions from the success of French bayonet advances against them in Italy in 1859: at Magenta and Solferino, the poorly trained and badly led Austrian infantry had been unable to draw much benefit from their technically advanced although still muzzle-loading, rifles; there had been inadequate training in range-fighting and sighting and, as a consequence, the French were able to close and use their bayonets.

The armies and navies of 1830 were still in most respects fighting like their late 17th century predecessors, with close-packed formations of troops on land using low-accuracy, slow-firing weapons at short range, and wooden warships totally reliant on wind power at sea, but the armies and navies of 1880 seemed, and in many respects were, very different. Changes in firepower, and the consequences of these changes, were each very important to this transition.

Steamships

'I sell here, Sir, what all the
world desires to have – Power.'

MATTHEW BOULTON ON THE STEAM ENGINE INVENTED BY
JAMES WATT WHICH HE PRODUCED AT HIS FACTORY

I N HIS *ON NAVAL WARFARE UNDER STEAM* (1858), GENERAL SIR
HOWARD DOUGLAS CLAIMED 'THE EMPLOYMENT OF STEAM AS
motive power in the warlike navies of all maritime nations, is a vast and
sudden change in the means of engaging in action in the seas, which must
produce an entire revolution in naval warfare'. Wind and wood, the
dynamics and properties of which had dominated naval power for
millennia, became redundant within decades in what truly was a total
revolution in means and capability.

The first steam warship, *Demologos* (Voice of the People), later renamed *Fulton*, was laid
down – when construction began – in 1814, in order to provide defence against British naval
power. It was intended for the protection of New York harbour during the war of 1812–15
between the two powers, but the war ended before it could be used.

Although the Americans were first, the British rapidly developed a steam capability at
sea. Their industrial capacity ensured that, even when other countries took the lead in
technological innovation, Britain would be best placed to catch up and to develop what
seemed appropriate.

STEAM WARSHIPS

The British navy at first leased small private steamships for use as tugs, but, in 1821, the tug
Monkey was bought, providing the navy with its first owned steamship. In 1822, the *Comet*,
the first steamer built for naval service, followed. Again it was a small ship, brig-sized and for
use as a tug, reflecting the manoeuvrability provided by steam power. The expedition sent
against Algiers in 1824 included HMS *Lightning*, a steam-powered paddleship equipped with
three guns, launched at Deptford in 1822. This was the first operational deployment of a
British steamship. Four British-built paddle steamers were used for the new Greek fleet from
1827, and they encouraged interest in steam power among serving British officers. HMS
Columbia, which entered service in 1829, was another armed steamship in the British navy,
and, the following year, the first purpose-built steam warship, HMS *Dee*, entered service.

Early steamships suffered from slow speed, a high rate of coal consumption, and the
problems posed by side and paddle wheels. These included their vulnerability to fire and the
space taken up by wheels and coal-bunkers: the wheels ensured that steamships carried few
guns (the *Columbia* only had two), so that, had it come to conflict with a ship of the line, it
was thought that the outgunned steamer would be sunk. There was also a reluctance to
throw away the existing British lead in sail-warships by embracing the new technology, a
frequent dilemma when dominant powers were faced by new opportunities.

HAZARDOUS WATERS

However, steam power replaced dependence on the wind, making journey times more
predictable and quicker, which was a requirement for both warships and merchantmen, and
increased the manoeuvrability of ships, making it easier to sound inshore and hazardous
waters, and to attack opposing fleets in anchorages. The ability to operate inshore was
shown in 1840 when the British fleet bombarded the Egyptian-held fortified port of Acre.
This was crucial to the evacuation of Acre by the Egyptians and to halting the Egyptian
advance into Syria.

The ability of ships to function in rivers and during bad weather was also enhanced by
steam power, as demonstrated in the First Burmese War of 1824–6 when the 60-horsepower

engine of the East India Company's steamer *Diana*, a 100-ton paddle tug built in India in 1823, allowed her to operate on the swiftly flowing Irrawaddy River. The *Diana* towed sailing ships, which destroyed Burmese war boats, and was crucial to the successful British advance 400 miles up-river, which led the Burmese to negotiate and accept British terms.

Although the British expedition that sought to travel up the River Congo in 1816 on the *Congo*, the first steamship on an African river, failed, due to problems with the boat, as well as disease and the difficult cataracts on the river, steam power was to transform river warfare and the capability of river-based forces. The French, for instance, expanded their strength in the Senegal Valley in West Africa from 1854, developing an effective chain of river forts linked by steamboats.

AMERICAN CIVIL WAR

Steamboats also played a major role on the inland waterways of the USA during the American Civil War (1861–5), more particularly in the Mississippi valley system. Union boats on the Cumberland and Tennessee rivers were important in the critical campaign in central Tennessee in February 1862, although, that year, the well-fortified Confederate position at Vicksburg was not reduced when bombarded from the Mississippi by Farragut's Union fleet. The Russians used steam-driven launches for their crossing of the Danube in 1877 in the face of Turkish opposition. The Dutch used steam gunboats in Borneo, the Portuguese on the Limpopo and Zambezi rivers in Africa, and the French too when they advanced down the River Niger on Timbuktu in 1894. They were also used by the British in their successful advance up the River Nile on Khartoum in Sudan in 1898 and, subsequently, five were utilized when the British moved further up the Nile in order to thwart French interest in southern Sudan.

On the high seas, steamships showed that they were able to cope with bad weather. The *Nemesis*, a 700-ton British iron-hulled paddle steamer, built at Birkenhead by John Laird for the East India Company, sailed through the winter gales off the Cape of Good Hope to China in 1840 and was the first such warship to reach Macao, although two lesser warships had crossed the Pacific from Chile the same year. Britain was at war with China in the First Opium War and *Nemesis* went on to destroy 11 Chinese war junks in January 1841 near Canton.

SCREW PROPELLER

In the 1840s, the screw propeller (placed at the stern) offered a better alternative to the paddle wheel, by making it possible to carry a full broadside armament, and this made the tactical advantages of steam clear-cut; screw steamers were also more mobile. The sloop *Rattler*, launched in 1843, was referred to in 1845 by Sir George Cockburn, an experienced admiral, 'The proof we have lately had of the efficiency of the screw as a propeller on board the *Rattler*... with all these efforts and improvements in continued progress, in which we are decidedly taking the lead, and are therefore in advance, I feel very confident.' The *Rattler* was followed in 1846 by the frigate *Amphion*, and the British quickly followed suit when the French ordered the *Napoléon*, the first screw ship of the line, converting *Ajax* to screw propulsion that year. Alongside new ships, there were conversions, encouraged by cost factors. The American fleet that covered their landing in 1847 in the Mexican–American War included a screw sloop, a paddle frigate and three leased paddle steamers.

While, by the start of 1854, France had the *Napoléon* and eight conversions, the British, who outspent them heavily in 1848–51, had constructed three new screw ships of the line and converted another seven. The British warships sent to the Baltic in 1854 against Russia in the Crimean War included no fewer than nine steam battleships, two designed as steam battleships and the other seven converted from sailing ships of the line, as well as six sailing ships of the line and four 60-gun blockships. Another four, plus four screw battleships and ten sailing ships of the line followed later in the year, although the fleet achieved little in the Baltic, largely because the Russians did not choose to send their fleet to sea. *Victoria*, launched in 1859, the largest wooden screw warship built, mounted 131 guns and cost £150,000.

NATIONAL CONFIDENCE

Steam also, however, created a British fear of attack that Viscount Palmerston spoke of in Parliament when he referred to a 'steam bridge' across the Channel able to serve the cause of French invasion. But, supported by an unparalleled network of coaling stations, steam also gave the British navy a range that led to national confidence in its power, as was noted by the fictional Captain Sir Edward Corcoran in Gilbert and Sullivan's comic operetta *Utopia Limited* (1893).

> *I'm Captain Corcoran, KCB*
> *I'll teach you how we rule the sea,*
> *And terrify the simple Gauls;*
> *And how the Saxon and the Celt*
> *Their Europe-shaking blows have dealt,*
> *With Maxim gun and Nordenfeldt*
> *(or will, when the occasion calls).*
> *If sailor-like you'd play your cards.*
> *Unbend your sails and lower your yards,*
> *Unstep your masts – you'll never want 'em more.*
> *Though we're no longer hearts of oak,*
> *Yet we can steer and we can stoke,*
> *And thanks to coal, and thanks to coke,*
> *We never run a ship ashore.*

Overseas British naval bases spanned the world. In 1898, they included Wellington, Fiji, Sydney, Melbourne, Adelaide, Albany, Cape York (Australia), Labuan (North Borneo), Singapore, Hong Kong, Weihaiwei (China), Calcutta, Bombay, Trincomalee, Colombo, the Seychelles, Mauritius, Zanzibar, Mombasa, Aden, Cape Town, St Helena, Ascension, Lagos, Malta, Gibraltar, Halifax, Bermuda, Jamaica, Antigua, St Lucia, Trinidad, the Falklands and Esquimalt (British Columbia). Nearly as far-flung, the French naval bases included Martinique, Guadeloupe, Dakar, Libreville, Diego Suarez, Obok, Saigon, Kwangchowwan, New Caledonia and Tahiti.

Steamships were also adopted by non-Western powers, albeit initially less successfully. In Vietnam, for example, the Minh-mang (1820–41) tried but failed to build steamships. However they remained crucial to the articulation of Western power in the 19th century.

Minié Bullet

'As the Russians come within 600 yards, down goes that line of steel in front, and out rings a rolling volley of Minié musketry. The distance is too great. The Russians are not checked, but still sweep onwards...'

WILLIAM HOWARD RUSSELL AT THE BATTLE OF BALACLAVA, 1854

THE HIGHLY SUCCESSFUL MINIÉ BULLET REPLACED THE MUSKET BALL and helped ensure that the rifle was the infantry weapon of choice. Developed by Captain Claude-Etienne Minié, from an idea by Gustave Delvigne and patented by Minié in 1849, this cylindro-conoidal lead bullet expanded when fired to create a tight seal within the rifle, thus obtaining a high muzzle velocity: the bullet contained an iron plug in its base and was cast with a diameter slightly less than that of the gun bore, a process helped by the greased grooves that ran horizontally around the base of the bullet. When fitted in the muzzle, it slid easily down the bore. When the gun was fired, the charge pushed the iron plug into the base of the bullet, causing it to expand and grip the rifling of the bore. Thus it was fired on an accurate trajectory. The combination of ease of loading and accuracy was a major advance that encouraged the mass use of rifled small arms.

A later form of the Minié bullet, improved by James H. Burton, an armourer at Harper's Ferry in the US, had a hollow base, which had the same effect of expanding when the charge went off and sealing the bore. This did away with the need for the iron plug. The charge was fired by an external percussion cap, and this system married up the reliability of fire of the latter with the greater accuracy of the Minié bullet. This was also a less expensive bullet.

RAPID FIRE

Thanks to its easy loading, the Minié bullet made for rapid fire and accuracy, and the Minié rifle was adopted by the British army in 1851. The Pattern 1851 Rifle Musket was then superseded in 1853 by the Pattern 1853 Rifle Musket, which was used with great effect in the Crimean War (1853–6). The combination of the Minié bullet and the percussion-lock rifle was deadly. The effective range of infantry firepower increased, and the casualty rates inflicted on close-packed infantry rose dramatically. Attacking Russian columns, seeking to close to bayonet point, took major losses from the Enfield rifles of the British at the Battle of Inkerman (1854). The formations and tactics of Napoleonic warfare, the column attacks and bayonet tactics employed successfully by the Austrians against the Piedmontese in 1844–9 at Custoza, Santa Lucia and Novara, now seemed likely to succeed only at the cost of heavy casualties.

In the American Civil War, massed frontal attacks on prepared positions became more costly and successful, as the Union discovered at Second Manassas (1862) and Fredericksburg (1862), and the Confederates at Corinth (1862), Stones River (1862–3), Gettysburg (1863) and Franklin (1864).

The improvement of firearms greatly affected practice and tactics, not least by making it easier for soldiers to fire from a prone position.

Minié bullets fired by percussion-lock rifles offered a much greater accuracy than balls from smooth-bore muskets, especially as the range increased. The Model 1855, the standard infantry weapon in use in the American army in 1861, fired the Minié bullet and had a muzzle velocity of nearly 300 metres (950 feet) per second.

The Minié bullet was part of a continual process of innovation in 19th-century firearms. This was an innovation in which wars made it possible to test the effectiveness of different weapons. As a result of the war between Austria and Prussia in 1866, for example, breech-loading rifles became seen as far better than the slower muzzle-loaders. In 1866, the French adopted the *chassepot* rifle, named after its inventor Antoine Chassepot, the head of the arsenal of Châtellerault. Invented in 1863, this rifle had a more gas-tight breech and a far greater range than the Prussian 'needle' rifle, which no longer represented cutting-edge military technology. In the Franco–Prussian War of 1870–1, the *chassepot* showed its value against Prussian frontal attacks. Whereas the 'needle' rifle fired a detonator located in the centre of the cartridge, the *chassepot* fired the charge by penetrating the propellant cartridge at its base.

THE EMPTY BATTLEFIELD

The improvement of firearms greatly affected practice and tactics, not least by making it easier for soldiers to fire from a prone position. The Prussian tactic of concentrating strength on the skirmishing line, and adopting more extended formations that were less dense than columns or lines, and therefore less exposed to fire, was seen as worthy of imitation. This reflected a tactical adaptation to new technology that represented an end to Napoleonic warfare showing us that firepower had an impact on battle techniques prior to the mass introduction of the machine gun. This looked toward what was to be termed the 'empty battlefield', as units sought to avoid exposure.

More generally, the Prussians in the war of 1870–1 proved superior in command and control, and were also helped by a fixity in purpose and a clearly planned strategy. Whereas the French essentially sought to muddle through, the Prussians operated a co-ordinated command system in which staff officers had a major role, with officers from the General Staff expected to advise commanders. This helped the management of risk and error. So also, at the micro-level, did an emphasis on professionalism, including the training of officers to take their own decision at all levels within the constraint of the command plan. A dynamic interaction between hierarchy and devolved decision-making meant that small unit operations supported and harmonized with those of large forces.

Rapid Prussian victories near the French frontier, culminating in the envelopment of French forces that surrendered at Sedan, were followed by a Prussian advance on Paris, and also across much of northern France. Defeated and heavily divided, the French accepted terms that included reparations and the loss of territories. This embittered relations between France and Germany and, in the long term, helped lead to World War I.

ALFRED NOBEL

Innovation after the Franco–Prussian War of 1870–1 included the adoption of smokeless powder, which burned more efficiently and permitted an increase in the range and muzzle velocity of bullets. Alfred Nobel, who instituted the Nobel Prizes in his will, perfected this power on the basis of a use of nitrocellulose, which had been discovered in 1846, and which ensured that a firearm propellant was available to replace gunpowder. Nobel's work in the late 1880s was speedily followed in the 1890s by the adoption of the powder.

There was also the development of an efficient system of magazine feed, permitting reliable repeating rifles using spring action to feed cartridges at a rapid rate. Two of the other major advantages of smokeless powder were that the field of vision of infantrymen was not now blocked by their own fire, and that the enemy had a harder time figuring out where fire was coming from.

Among the clip-fed breech-loading rifles were the French Lebel (1886) and the German Mauser (1889). Having moved from muzzle-loading Enfield rifles to Sniders (the British army's first breech-loading firearm) and then Martini-Henrys (adopted in 1871), the British switched to Lee-Metford magazine rifles. Invading Ethiopia in 1868, the Ethiopians were outgunned, their matchlocks and smooth-bore shotguns less effective than the British rifles, whether muzzle-loading or the more recent breech-loaders. British volley fire proved devastating. The British indeed were defeated, with heavy losses, in Afghanistan at Maiwand in 1880 by opponents armed with British Enfield rifles. In 1895, the Portuguese defeated the kingdom of Gaza in southern Mozambique: in battle Portuguese squares used their Kropatshek magazine rifles to overcome Gaza charges. In Sudan, in 1898, where the British fought the Mahdists, a company of 100 British troops could fire ten shots per yard per minute across its 100-yard front if armed with Martini-Henry or Remington rifles which were effective from 1,500 yards. Martini-Henrys fired a heavy bullet and this was very useful against attacking forces.

The increased velocity of rifles made them more deadly and the development of the spitzer or boat-tail bullet proved a smaller, more aerodynamically stable bullet.

SUPERIOR MARKSMAN'S AID

In the Boer War (1899–1902), in southern Africa, the British suffered from the superior marksmanship of their Boer opponents with the smokeless, long-range Mauser magazine rifles. In response, the British developed an appropriate use of cover, creeping barrages of continuous artillery fire, and infantry advances in rushes, co-ordinated with the artillery. This was important to their success, but the massive resources of the British empire were far more so. Technology played a major role in their application, as steamships ensured that large numbers of troops could be moved to South Africa, and the railway then enabled them to be moved into the interior. There were troops not only from Britain but also from the empire, including Australia and Canada.

The Boers were not alone among Britain's opponents in having effective rifles. Donald Alexander MacAlister, who served with the British field force sent against the Aros of south-eastern Nigeria, recorded in 1902, 'The natives have been crowding in with guns of all kinds and some of these guns are very fine specimens of Sneider. The bulk of those brought in lately could if properly handled have done us a good deal of damage. They have all been broken up and burned.'

Railways

'General Sherman has given orders for
the farther destruction of all public
property in the city... the arsenal,
railroad, depots... there is not a rail upon
any of the roads within twenty miles of
Columbia but will be twisted into
corkscrews before the sun sets.'

GEORGE NICHOLAS REPORTING ON THE AMERICAN CIVIL WAR

RAIL TRANSFORMED THE POSSIBILITIES FOR LAND WARFARE in the 19th century, providing key operational and strategic advantage. In contrast, train-mounted weaponry was of limited consequence. On 27 October 1883, Albert Robida published in *La Caricature* an anticipation of warfare in the 20th century that included electric-powered, armoured trains firing cannon and machine guns. Armoured trains indeed played a part, for example in the Russian Civil War and in the Chinese civil wars of the 1920s, but there was no equivalent to the use of the internal combustion engine.

Operationally, however, railways played a vital role in the mobilization and deployment of forces. They enabled troops to be moved rapidly to the battlefield. Here there was a major contrast, between the Crimean War in 1853–6, in which the lack of Russian rail links to the Crimea, the crucial sphere of operations on land, hit their deployment and logistics, and the Austro-French war in Italy in 1859 in which both sides employed railways in the mobilization and deployment of their forces. In the opening stage of the war, the French moved 50,000 troops to Italy by rail, thereby helping to gain the initiative.

CHANGING THE COURSE OF THE AMERICAN CIVIL WAR

In the American Civil War (1861–5), the railway, which had expanded greatly in the 1840s and 1850s, especially in what became the North, also made a huge difference. It helped the Union mobilize and direct its greatly superior demographic and economic resources, and played a major role in particular battles, and did so from the outset. Reinforcements under Joseph Johnston arriving by train on the Manassas Gap Railroad, helped the Confederates win at First Manassas/Bull Run (1861). The following year, the Confederate commander Braxton Bragg was able to move his troops 776 miles by rail from Mississippi to Chattanooga, and thus create the opportunity for an invasion of Kentucky. Rail junctions, such as Atlanta, Chattanooga, Corinth and Manassas, became strategically significant, and the object of operations. In turn, operational plans depended on using or threatening rail links, as in the campaign that led to the Battle of Second Manassas/Bull Run (1862), when 'Stonewall' Jackson hit the Union supply route along the Orange and Alexandria Railroad, destroying the supply depot at Manassas Junction. Later that year, the Union commander Ambrose Burnside planned to move south along the Richmond, Fredericksburg and Potomac Railway after he had captured Fredericksburg, which in the event, he failed to do.

At the tactical level, man-made landscape features created for railways, such as embankments, played a part in battles. Jackson concealed his men in an abandoned cutting for an unfinished railroad at Second Manassas/Bull Run, and subsequently used it as a defensive position. The Union's dependence on railways led to the Confederacy raiding both them and the telegraph wires that attended and controlled them, as with John Morgan's raids in Tennessee and Kentucky in 1862. Similarly, in 1864, Union cavalry in the Atlanta campaign raided the Atlanta and West Point, and Georgia Railroads. After a raid on the Macon and Western Railroad failed, Sherman, the local Union commander, moved his army to cut the railway, and forced the Confederates to abandon Atlanta.

The new potential for transport and control created the problem of requiring revised standards of administration. The Union created US Military Railroads as a branch of the War Department, although this did not fully exploit the powers of the president under the 1862 Act, which gave him the authority to take control of the railways. US Military Railroads also

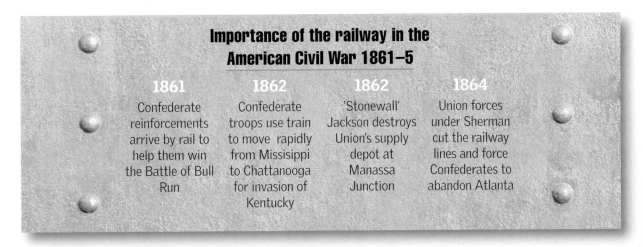

Importance of the railway in the American Civil War 1861–5

1861	1862	1862	1864
Confederate reinforcements arrive by rail to help them win the Battle of Bull Run	Confederate troops use train to move rapidly from Missisippi to Chattanooga for invasion of Kentucky	'Stonewall' Jackson destroys Union's supply depot at Manassa Junction	Union forces under Sherman cut the railway lines and force Confederates to abandon Atlanta

built and repaired track and bridges, providing a quick-response system to enhanced transport links as the exigencies of the war required.

In its wars with Austria in 1866 and France in 1870, Prussia won in part thanks to an effective exploitation of the railway network to achieve rapidly the desired initial deployment, thereby gaining strategic initiative. The Crimean (1853–6) and Franco–Austrian (1859) Wars had led to an understanding of the need for good rail links. Through use of planned rail movements, the Prussians made mobilization a predictable sequence and greatly eased the concentration of forces. This was of particular value in a short conflict, such as their wars with Austria (1866) and France (1870–1). Uniform speed by all trains was employed by the Prussians in order to maximize their use. Military trains were also of standard length. Rail use was planned and controlled by a railroad commission and by line commands, creating an integrated system linked and made responsive by the telegraph. The number of locomotives was kept high in order to cope with a wartime rise in demand. After their defeat in 1866, the Austrians understood these advantages, and Archduke Albert, who became Inspector-General of the Austrian army, also improved rail transport.

IMPERIAL EXPANSION

Railways were also important on the very different scale of imperial expansion and consolidation outside Europe. Communication links were sought by imperial powers and were seen as a way to strengthen their empires politically, economically and militarily. Faced by the Métis' (mixed-blood population) rebellion in Manitoba and Saskatchewan in 1885, the Canadian government sent over 4,000 militia west over the Canadian Pacific Railway, achieving an overwhelming superiority that helped bring victory. This was followed by a new increase in government subsidy for the railway that enabled its completion that year. The role of the railway was also strategic in the broadest sense in that it helped in the development of economic links that sustained and strengthened key powers. For example, the railway was crucial in developing economic links between coastal and hinterland America, as well as in integrating the frontiers of settlement with the world economy. This was important both in the spread of ranching, with the cattle being driven to railheads, and of mining.

In 1896, the Russians obliged China to grant a concession for a railway to Vladivostok across Manchuria, and this Chinese Eastern Railway was constructed in 1897–1904. Also in 1896, the British army under Kitchener invading Sudan built a railway straight across the desert from Wadi Halfa to Abu Hamed, the 383-mile-long Sudan Military Railway constructed across the bend of the Nile. It was pushed on to Atbara in 1898, and played a major role in the supply of the British forces. By 1900, the British had built 10,000 miles of railways in India, and rail links came there to play a greater role in geopolitics, strategy and logistics. In 1897, the British in India moved troops by train against the Waziris on the North-West Frontier. Three years later, in China in the Boxer Rising, the destruction of part of the track between Tientsin and Beijing forced the abandonment of the initial attempt to relieve the foreign legations in the capital. On the other side of the Pacific, the building of a railway across the rebel area in 1900 helped to end longstanding Mayan resistance to the Mexicans, although the effects of cholera, smallpox and whooping cough were also significant. In the Boer War (1899–1902), the railways that ran inland from the ports facilitated the deployment of British military resources, although wagon trains also proved necessary. Fear of an invasion of Canada during the Anglo-American Venezuela crisis in 1904 led to surveillance of nearby American railways. In Russia, the capacity of the Trans-Siberian Railway, and its uncompleted section round Lake Baikal, were important in Russia's attempt to defeat Japan in the war of 1904–5: the railway could move about 35,000 men a month. The Russians were determined not to repeat their failures of the Crimean War.

Road construction was also crucial, as in New Zealand in the 1860s, when British troops were used to extend the Great South Road from Auckland over the hills south of Drury in the face of Maori opposition. By 1874, a road had been finished between Tauranga and Napier, separating two areas of Maori dissidence.

Railways were of less direct use in battlefield operations – troops could not move on the battlefield – and in the Third World did not play a role in many areas. For example, no railways were built in Afghanistan, a country that was never brought under European control. Horses, mules and oxen provided essential mobility for long-range movements away from rail links.

CONFLICT ZONES

Railways remained vital during the 20th century, being the major way in which troops and supplies were moved to conflict zones in World War I. The trench warfare of World War I depended on the combatants' ability to provide large quantities of supplies on a regular basis in order to support the densely packed forces necessary to maintain a continuous front line and the large number of guns.

The contrast was readily apparent where railways were sparse, as in Africa during World War I and China in the 1920s. In these cases, the communication system was heavily dependent on human porters.

Despite the spread of petrol-driven vehicles, railways remained important in World War II, and this helped explain the emphasis placed upon bombing them. This was particularly seen in the Allied air assaults on Germany and Japan, which identified rail links as fundamental to the economies of the combatants. Furthermore, rail links were seen as operationally crucial. Thus, prior to Operation Overlord, the Allied invasion of Normandy in 1944, Anglo-American air power was launched in a major effort to isolate the zone of hostilities, particularly by cutting rail bridges across the Seine and the Loire in order to weaken German re-supply capabilities.

Rifled Artillery

'Now artillery has changed everything. A cannon ball knocks down a man six feet tall just as easily as one who is only five feet seven. Artillery decides everything, and infantry no longer do battle with naked steel.'

ARMIES ALL OVER EUROPE REAPED THE BENEFITS OF THE REPLACEMENT OF SMOOTHBORE BY RIFLED ARTILLERY which, as with rifles (see p. 102) greatly increased their effectiveness. Rifled steel artillery was difficult to produce because it was necessary to find ways to cool barrels evenly in order to prevent them from cracking. Alfred Krupp's improvements in casting methods opened the way in the 1840s, and in 1851 he showed an all-steel cannon at the Great Exhibition in London thereby earning the nickname of 'the Cannon King'. Stronger, and more durable, than bronze guns, steel guns were able to take larger charges and therefore offered greater range.

Rifled artillery allowed the cannon to move back out of the range of accurate infantry fire, without losing their own accuracy. It showed its range and effectiveness at the expense of the Russians in the Crimean War, both at the Battle of the Alma in 1854, and at Traktir on the River Tchernaya in 1855. In the latter, explosive shells fired by French rifled cannon caused heavy Russian casualties among the force that unsuccessfully sought to dislodge the Anglo-French army besieging the Russian base of Sevastopol.

The rifled artillery also had a major impact in the hard-fought battles of the Franco-Austrian War of 1859. The new French rifled cannon were superior to their Austrian smoothbore counterparts and destroyed most of them with highly accurate counter-battery fire, before devastating the Austrian infantry, as at the Battles of Magenta and Montebello. At Magenta, General MacMahon pushed forward his guns to provide vital cover for the advancing French infantry. The effectiveness of artillery was also enhanced by optical sights. Although developed in the mid-17th century, they only became common in the mid-19th.

PRUSSIAN TACTICS

In the Danish-Prussian War of 1864, the Danes suffered because they lacked rifled artillery. Similarly, in the American Civil War, artillery remained dominated by muzzle-loaders firing solid shot by direct fire. In the Austro-Prussian War of 1866, however, the Austrians outranged the Prussians and hit their infantry advances, although their small-unit character minimized casualties. The Austrians had introduced effective rifled artillery after their exposure to the French in 1859. After 1866 the Prussians improved their artillery tactics and organization, and pushed their guns to the front to make them more telling in battle.

In the Franco-Prussian War of 1870–1, the Prussians were superior to the muzzle-loading French artillery. Ironically, the French had turned down an approach from Krupp to sell them the breech-loading steel artillery he made for the Prussian government. At the early Battles of Fröschwiller and Wörth, the Prussians benefited from their improved artillery. The French entrenchments proved an easy target, and the use of percussion, rather than time-fused, shells made it easier to ensure accurate fire.

At Mars-la-Tour, Prussian artillery repeated its decisive role, helping to force the French to fall back to Metz. Similarly, at Sedan, Prussian artillery in the surrounding hills provided a good position for bombardment, while, when the French tried to gain the hills, they were driven back by heavy and accurate artillery fire. In 36 hours at Sedan, the remorseless Prussians fired 35,000 shells. Had the French had artillery to compare then the Prussian attacks, especially at Wörth and Gravelotte would have been devastated. The difference was to be seen in the German attacks in 1914 and, in particular at Verdun in 1916, when the Germans encountered modern weaponry.

COLONIAL EXPANSION

Rifled artillery was also used to further Western colonial expansion. In Senegal and Algeria, the French used artillery to breach the gates of fortified positions, and then stormed them. The Treuille de Beaulieu rifled mountain gun, a mobile light mortar, was first tested by the French in Algeria in 1857. Artillery, especially 95 mm siege guns using more powerful explosives, played a crucial role in the conquest of the Tukulor forts in the Sahara by the French in 1890–1. That decade, the Belgians used Krupp artillery to help overcome opposition in the Congo, while, further east, in East Africa, the Germans used their Krupps against the Umyamwezi: the latter's rifles were outgunned. In Sudan in 1898, advancing British troops at Atbara outgunned the Mahdists, who had no artillery. Five years later, the walls of Kano in Nigeria were breached within an hour by British cannon. A British intelligence report on the fighting in Somaliland, where the British invaders were engaged against a fundamentalist Islamic movement, noted, 'It will also be necessary to employ artillery, firing high-explosive shells, if the various Dervish strongholds are to be captured without very heavy casualties.'

Problems arose for the imperialists, however, when non-Western forces used such weaponry. At Maiwand in 1880, in the Second Anglo-Afghan War, the Afghans successfully employed Armstrong guns against the British, while in 1900, the foreign legations besieged by the Boxers in Beijing were put under pressure from the Chinese use of a Krupp's quick-firing cannon.

> As an example of the weight of firepower available, the Italians, not the strongest or most industrialized of powers, deployed 1,200 guns for their attack on the Austrians in the Third Battle of the Isonzo in October 1915. The British used 2,879 guns – one for every nine yards of front – for their attack near Arras in April 1917, far more than those they had used when launching their offensive on the Somme on 1 July 1916.

Thanks to mass-production, the new artillery could be adopted in large quantities. The availability of steel weaponry owed much to improvements in production methods, especially the Bessemer steel converter and the Gilchrist-Thomas basic steel process. As a result, output of steel rose dramatically from the 1870s. In 1900, Russia ordered 1,000 quick-firing field guns from the Putilov iron works, while between 1866 and 1905, Prussia doubled the complement of field guns to 144 per infantry corps.

GREATEST KILLER

In World War I, the greatest killer on the battlefield was artillery – possibly causing up to 70 per cent of battlefield deaths, followed by machine-guns and rifles. 58.51 per cent of British battlefield casualties were caused by high-explosive shells and trench mortar bombs compared to 38.91 per cent by small arms fire. The French 75 mm field-gun could fire over 15 rounds a minute and had a range of 9,000 yards, while German 150 mm, field howitzers could fire five rounds per minute. Air-burst shrapnel shells increased the deadly nature of artillery fire against which the spread of steel helmets offered little protection. Furthermore, the relative stability of the trench systems made it worthwhile deploying heavy artillery to bombard them. The guns could be brought up and supplied before the situation changed, whereas this was not possible in manoeuvre warfare. It was also necessary to provide artillery support to batter an enemy's defensive systems.

As artillery came to be seen as the method to unlock the static front, so its quantity greatly increased. As an example of the weight of firepower available, the Italians, not the strongest or most industrialized of powers, deployed 1,200 guns for their attack on the Austrians in the Third Battle of the Isonzo in October 1915. The British used 2,879 guns – one for every nine yards of front – for their attack near Arras in April 1917, far more than those they had used when launching their offensive on the Somme on 1 July 1916. Then the guns had been spread too widely to be effective.

During the war, the number, strength, precision and use of artillery improved vastly. The French were very much helped in resisting the Germans at Verdun in 1916 by an effective use of artillery, now heavier than their guns in 1914. This included the creeping barrage when gunfire falls just in front of advancing troops, employed to support the counter-offensive in October 1916. On 18 July 1918, the French counter-offensive on the Marne was supported by a creeping barrage, with one heavy shell per 1.27 yards of ground and three field artillery shells per yard. The British made effective use of artillery-infantry co-ordination in the final victory of the Western Front in 1918.

KEY ELEMENT

In place of generalized firepower, there was systematic co-ordination, reflecting precise control of both infantry and massive artillery support, and improved communications. The British army had 440 heavy artillery batteries in November 1918, compared to six in 1914, and inflicted considerable damage on German defences. Counter-battery doctrine and tactics had developed appreciably and the British had become adept at indirect (three-dimensional) firepower.

In World War II, the Allies had important advantages in artillery, and these were particularly significant because, as in World War I, more battlefield casualties were killed by artillery than by any other weapons system. Artillery was more deadly than in the earlier war because of better shells and fuses, for example proximity fuses, which were used by the Allies in land warfare during the Battle of the Bulge in December 1944. Benefiting from impressive guns, such as the American 105 mm howitzer, Allied artillery was more intensive in overwhelming firepower, although the British lacked an adequate modern heavy artillery. The British, Americans and Soviets were very keen on using big artillery bombardments to accompany their offensives (the Soviets had particularly plentiful artillery), and the Germans, who used artillery when they could, had no real answer. In January 1945, the Soviets were victorious in the Vistula-Oder offensive, fighting their way through the German defence lines, greatly aided by their plentiful artillery, in which their margin in numbers was about 7.5 to 1. The Japanese, who lacked an artillery to compete with American firepower, relied on the terrain, frequently digging in underground.

Artillery continues to this day to be a key element in winning battles. It was important to the British overland advance against Argentinian positions in the Falklands War in 1982 and has also played a significant role in recent conflict in the Middle East, being used, for example, by the Israelis in bombarding Hizbollah positions in Lebanon in 2006.

Ironclads

'He who commands the sea has

IRONCLAD SHIPS ARE BELIEVED TO HAVE BEEN USED BY THE KOREANS in the 1590s in conflict with Japan. These 'turtle ships', were oar-driven boats, possibly covered by hexagonal metal plates in order to prevent grappling and boarding, and they were used when the Japanese fleet was defeated at the Battle of the Yellow Sea in 1592.

This tradition, and, more generally, East Asian naval effectiveness, however was not developed further. Instead, the genesis of ironclads lay with developments in naval ordnance, particularly the introduction in the 1820s of exploding shells in place of solid shot. Colonel Henri-Joseph Paixhans, who developed the necessary type of gun, pressed for the combination of his new ordnance, adopted by France in 1837, with steamship technology. The new ordnance posed a terrible threat to wooden ships, and the French hoped that it would enable France to threaten British naval hegemony.

Shells helped lead to their antidote, armoured warships. The first iron-clad warship was French and was inspired by the success of three newly built French 1,575 ton iron-plated, wooden floating gun platforms (which had been towed from France by steamships) off Kinburn in 1855 during the Crimean War. Laid down in March 1858, and launched in November 1859, the 5,630-ton *La Gloire* was merely a wooden frigate fitted with 4.5 inch-thick metal plates, since the French were handicapped by a lack of ironworking facilities. Nevertheless, Napoleon III, who was determined to challenge the British, ordered five more ironclads in 1858 and they were commissioned in 1862.

THAMES IRONWORKS

Worried that their naval lead was being destroyed, the British matched *La Gloire* with an armoured frigate, HMS *Warrior*, laid down in May 1859 and completed in October 1861. With its iron hull and displacement of 9,140 tons, this was more significant than the *La Gloire*, and, built at the Thames Ironworks, also testified to British manufacturing capability. The *Warrior* could also make 14 knots, compared with the *La Gloire's* 13, and carry 200 tons more coal. This ensured that it had a greater range.

The *Warrior*, which cost £377,000 to build, was a revolutionary ship design. It was actually a true iron ship, as opposed to an ironclad ship. Furthermore, with its watertight compartments below, it was the first large sea-going, iron-hulled warship.

The shift to iron reflected not simply the vulnerability of wooden ships to shellfire, but also success in overcoming the problems that had delayed the use of iron, including its effect on magnetic compasses, the fact that iron hulls fouled very much worse than those that were copper-bottomed, and the difficulties of securing sufficient consistency and quality in the iron. The ability to overcome these problems reflected the strength of the British economy in the acquisition and application of relevant knowledge.

Iron ships were structurally stronger, and made redundant the wooden screw steamers built in large numbers in the 1850s and the very beginning of the 1860s; Britain's last wooden screw ship of the line entered service in 1861.

BRITAIN'S LEAD

Having gained the lead in iron ships, the British retained it, ensuring a major advantage over the French navy. France's defeat by Prussia in 1870–1 confirmed the naval superiority already demonstrated by Britain in the ironclad building race of 1858–61, as it led France to focus on strengthening its army for a future war with Prussia.

Naval technology continued to change and to an extent that would have been inconceivable in the 18th century. The tension between armour and armament, weight and manoeuvrability, not least the mutually interacting need for more effective guns and stronger armour, led to changes in armour, and the iron, or in some cases in the 1870s the composite iron and wood navy was followed, after the introduction of compound armour plate in 1877, by the iron and steel navy. The British continued to make large hull components, like the keel, stern post and stern, out of iron until the 1880s, but there were also moves towards the first all-steel warships in the 1870s.

Earlier, the new steam-driven ironclad warships had been used against each other in the American Civil War. The inconclusive duel between the *Monitor* and the *Merrimack* (renamed the *Virginia* by the Confederates) in Hampton Roads on 9 March 1862 was the first clash between ironclads in history, although that was because the three European navies that already had commissioned ironclads (Britain, France and Italy) had not been at war with each other. In Hampton Roads, cannon shot could make little impact on the armoured

Ironclads

1853
The powerful guns of the Russian Black Sea fleet annihilate the Turks at the Battle of Sinop
1858
The first French iron-plated ship *La Gloire* is built
1859–60
HMS *Warrior* is built at Thames Ironworks at Blackwall
1862
The Confederate ironclad, the *Virginia* attacks the wooden warships blockading Norfolk, Virginia
1862
The first clash between two ironclads in the American Civil War at Hampton Roads
1866
At the Battle of Lissa, the Austrians inflict heavy losses on the Italians
1873
Germany now has the third largest armoured fleet in the world
1879-83
Superior Chilean warships overcome the Peruvians in the War of the Pacific

sides of the two ships even though they fired from within 100 yards. There was only one casualty in the engagement. The resilience of ironclads to cannon fire was indicated by the disparity in the Civil War between seven armoured Union ships lost to Confederate mines, compared to only one lost to fire from shore batteries.

The potential of ironclads had been shown on 8 March 1862, when the *Virginia* attacked the wooden warships blockading Norfolk, Virginia, employing ramming to sink one and using gunfire to destroy another. The *Virginia* itself was scuttled on 11 May 1862, as Union ground forces advanced.

During the war, the capability of ironclads increased. Whereas the *Monitor* had had two guns in one steam-powered revolving turret, the Union laid down its first monitor (a class of ship now taking the name of the original vessel invented by Captain Ericsson) with two turrets in March 1863, ultimately completing nine of the latter. One ship was converted to a three-turret monitor, the *Roanoke*, although it rolled too badly to be effective. It represented an advance in goals for the monitors, as it was designed for the high seas, whereas most monitors had too shallow a draught for this end.

IRONCLAD SUPREMACY

The Union also added iron armour to many of its ships. Some were 'tinclads', with only thin armour, but others, the 'city class' built at St Louis, had 6½ cm- (2½ inch) thick armour. The Confederate loss of New Orleans and Memphis in 1862 reduced its ability to build or convert warships for service on the Mississippi and other inland waterways, although the mobilization of the Confederate economy and the adaptability of the available manufacturing resources led to building ships at other river shipyards such as Selma and Shreveport. This permitted the construction of ironclads at places away from the coast that were not as vulnerable to capture by Union forces. However, it was quite an ordeal for the Confederates to get their ironclad *Tennessee* down the Alabama River from Selma to Mobile Bay. Both sides had ironclads when David Farragut's Union fleet successfully fought its way into Mobile Bay in August 1864. Four Confederate ships were defeated by 18 Union ships, and the fortifications at the Bay's entrance failed to block Farragut. The single Confederate ironclad was bombarded into surrender.

The Union also sent an ironclad, the *Camanche*, to San Francisco in order to protect California from Confederate raids. The ship was built in 1862–3 then divided into parts and shipped around Cape Horn – as yet there was no transcontinental railway. Once at San Francisco, it was reconstructed. In 1865–6, another ironclad steamed round Cape Horn to California. In 1865, the *Shenandoah*, the first composite (iron and wood) hulled cruising warship wrecked much of the New England whaling fleet in the northern Pacific. By the end of the war, the Union navy included 49 ironclads, although the inability to roll thick iron plates affected the effectiveness of their armour. This would have ensured that had Britain and the Union gone to war, as indeed appeared possible, the Union navy would have been vulnerable.

The use of ironclads in the American Civil War encouraged their construction in Europe: an example of diffusion between the European-American and European worlds. At Lissa (1866), the largest naval battle between Trafalgar (1805) and Tsushima (1905), the Austrians inflicted heavier losses on Italians in a confused mêlée of ship-to-ship actions. Investment in naval strength led to an expansion in fleet size in the 1860s and 1870s. The pace of German naval construction rose hugely in 1873 at Bismarck's insistence. By 1883, Germany had the third largest armoured fleet in the world.

Ironclads were also used further afield, for example in the War of the Pacific (1879–83) when, off Punta Angamos in 1879, armour-piercing Palliser shells, fired from the 9-inch Armstrong guns of Chilean warships, forced the badly damaged Peruvian *Huáscar* to surrender: the eight inches of wrought iron on its turret proved no defence. This battle was important because Chilean effectiveness in part depended on the ability to mount amphibious operations against Peru. The use of ironclads in this conflict reflected the wide-ranging impact of new naval technology.

Battleships

'A fully-equipped duke costs as much to
keep up as two Dreadnoughts... and
they last longer.'

DAVID LLOYD-GEORGE, 1909

AS THE GREAT POWERS IN THE 19TH CENTURY STROVE TO CONSOLIDATE THEIR POSITIONS they all took advantage of changes in naval technology. This led to the transformation of ships of the line into battleships, floating behemoths, not dependent on wind or weather, but able to deliver formidable quantities of firepower and also to take a heavy pounding. The pace of technological advance was very high, a key aspect of modern military capability. Completed by the British in 1861, HMS *Warrior* was a revolutionary ship design, the first large sea-going iron-hulled warship. There was also major progress in steam technology, with resulting fuel efficiency, speed, range and reliability. In the 1860s, high-pressure boilers were combined with the compound engine, and in 1874 the triple-expansion marine engine was introduced, although it was not used in warships until the 1880s. This engine was followed by the water tube boiler. By 1885, only distant-service ships still required sail, an equivalent to the use of cavalry in imperial warfare.

There were also major changes in armament and armour. The practice of locating heavy guns in an armoured casement began with the British HMS *Research* in 1864. The race between armour and armament, the problem of warship weight and manoeuvrability, led to a revolution in armour from the mid-1870s to the early 1890s, culminating in the nickel-steel plate patented in Germany by Friedrich Krupp from which warships were increasingly built. This gave added protection without added weight, encouraging the construction of bigger ships.

FRESH INVESTMENT

The naval force that could be applied was amply demonstrated at Alexandria in Egypt in 1882. Although many of their shells missed, 14 British warships with powerful guns, including HMS *Inflexible*, which had four 16-inch guns, inflicted great damage, with few British casualties: the shore batteries were not particularly well handled by the Egyptians, and the British warships did not have to face mines or torpedoes.

Launched in 1881, *Inflexible* was the first battleship fitted with vertical compound engines, as well as electric power for lighting, including the searchlights judged necessary to warn of attack by torpedo boats. However, these features led to its previously unmatched cost, £812,000, and to a long construction period of seven years. *Inflexible* was the prototype for four other British battleships laid down in the late 1870s.

The extent of new naval technology ensured that fresh investment was needed in order to avoid obsolescence. The 100-ton, nearly 44-foot long, Armstrong breech-loaders manufactured for HMS *Victoria*, which was launched in 1887, were the largest and most powerful guns in the world. In 1889, the British Naval Defence Act led to the expensive commitment to a two-power standard, so that Britain could be in a position to fight the second- and third-largest naval powers combined, a goal that arose from an apparently more threatening international situation. In this *fin-de-siècle* arms race, the British built nine battleships of the Majestic class, followed by the laying down, in 1898–1901, of 20 battleships modelled on the Majestic class. The eight battleships of the King Edward VII class followed, starting in 1902–4. HMS *Dreadnought*, which became the prototype of a faster and more heavily gunned type of battleship, was launched by the British in 1906. It was the first of a new class of all big-gun battleships, and the first capital ship in the world to be powered by the marine turbine engine.

The Japanese victory over the Russians at Tsushima the previous year had appeared to show what naval warfare would be: battles between battleships, with the capacity for total victory, and this victory leading to the end of the war. Six Russian battleships and two cruisers were sunk in the battle, four battleships were captured, and one battleship and three cruisers were scuttled to prevent capture. The 12-inch guns of the Japanese battleships inflicted the damage.

The *Dreadnought* made the earlier arithmetic of relative naval strength redundant, and encouraged the Germans to respond with the construction of powerful battleships: four were begun in the summer of 1907. Theirs was a naval race with Britain, and Germany built the world's second largest battle fleet. However, as the British were determined to stay ahead, and willing to pay to do so, German shipbuilding simply encouraged the laying down of more Dreadnoughts despite their high cost of approximately £1.7 million which was over 50 per cent more expensive than any previous battleship. The naval race continued because attempts in 1909–10 and 1912 to negotiate a treaty fixing the ratio of strength between both navies failed. By the outbreak of World War I in 1914, the British had 21 Dreadnoughts in service, the Germans 14, and 12 and five respectively were under construction. Germany's failure in this race ensured a shift to backing submarine warfare.

Other countries, including France and Russia, also swiftly built or ordered Dreadnoughts. The Americans laid down their first Dreadnought in 1906 and had four finished by the end of 1910, while Japan laid down their first two Dreadnoughts in 1909. Austria and Italy also engaged in a naval race.

BATTLE OF JUTLAND

Such ships were seen as crucial to great-power status, but in World War I the battleships played a smaller role than had been anticipated. The Battle of Jutland of 31 May–1 June 1916 between the British and German fleets, was not the hoped-for Trafalgar, nor a repetition of Tsushima. The Germans attempted to achieve a surface victory, by luring part of the British Grand Fleet with the main force of their High Seas Fleet (a goal they tried on several occasions), only to be confronted with the entire British fleet. However, the caution of Jellicoe, the British commander, concerned that German torpedoes would sink pursuing British ships, possibly denied the British the victory they might have obtained. The British lost more ships and over 6,000 men in the battle, but German ships were badly damaged in the big-gun exchange, and their confidence was hit by the resilience and size of the British fleet.

As the Germans finished not one of the Dreadnoughts or battle-cruisers they laid down during the war, compared with the five battle-cruisers laid down and completed by Britain, they did not have the margin of safety provided by a shipbuilding programme to fall back upon; and nor did they have the prospect of support from the warships of new allies that Britain benefited from -- with the entry of Italy into the war in 1915 and, more significantly, of the USA two years later. These additions more than nullified the success of German submarines in sinking Allied warships. The German failure in surface-ship warfare, and in the arithmetic of surface-ship strength, helped again to accentuate the importance for them of submarines.

As part of the peace settlement after World War I, Germany lost its fleet, while in 1922 a ratio in the capital ship tonnage was agreed by Britain, Japan and the USA in the Washington Naval Treaty, which also included an agreement to scrap many battleships and to stop new construction for ten years. Although some commentators argued that battleships were now obsolete in the face of air power and submarines, big surface warships had a continued appeal, and not simply for the European powers. Indeed, there was opposition to a stress on carriers becoming the key capital ships; and, in the 1930s, both the Americans and British put a major emphasis on battlefleet tactics based on battleships. So also did the Japanese: proponents of mass naval aviation met serious opposition from supporters of battleships. In the *Zengen Sakusen* (Great-All-Out Battle Strategy), the Japanese focused on the successive use of submarines, long-range shore-based bombers, carrier-based dive-bombers, and destroyer night-time torpedo attacks against the advancing American fleet, leading up to an engagement by the Japanese battleships. Under the Marusan Programme of 1937, they began to build the *Yamato* and *Musashi*, which were to be the most powerful battleships in the world. In the USA, keels were laid for four comparable 45,000-ton battleships in 1941, and seven were projected at over 60,000 tons each. The Germans built large surface warships once the peace settlement limitations on the German fleet were broken in 1935.

The role of battleships was enhanced by the absence of any major change in their design comparable to those in the late 19th century and the 1900s. With the arrival of the Dreadnoughts, battleship architecture reached a new period of relative stability, although there were considerable efforts to increase resistance to air attack. Armour was strengthened, outer hulls were added to protect against torpedo attack, and anti-aircraft guns and tactics developed.

A STRATEGIC IRRELEVANCE

World War II fully revealed the vulnerability of battleships to submarines and, even more, air power. The British HMS *Prince of Wales* and the HMS *Repulse* were sunk by Japanese planes in December 1941 when they tried to thwart the invasion of Malaya. Earlier that year, the German *Bismarck* was crippled by a hit on the rudder by a British aircraft-launched torpedo and this provided an opportunity for the British surface ships to come up and destroy the ship. In April 1945, the Japanese sent their last major naval force, led by the battleship *Yamato* to disrupt the American force that had landed on Okinawa, but it was intercepted by American bombers and sunk. The vulnerability of surface warships without air cover was amply demonstrated. The battleships on which the Japanese had spent so much had become an operational and strategic irrelevance. Nevertheless, if aircover existed battleships played a major role in shore bombardment, as on D-Day in 1944.

After 1945, the age of the battleship passed, as those built in the inter-war period were scrapped, to be followed by others launched in wartime. Commissioned in 1946, the British HMS *Vanguard* was the sole European battleship commissioned after the war, and was scrapped in 1960. In the Suez crisis of 1956, the French *Jean Bart* became the last European battleship to fire a shot in anger.

Machine Guns

'Whatever happens we have got
the Maxim gun, and they have not.'

Hilaire Belloc, 1898

THE MACHINE GUN WAS TO BE A MAJOR KILLER, but, at first, did not revolutionize the battlefield. Indeed, early use of the machine gun was limited. Colt's machine gun was used by the Americans in the Mexican War (1846–8) and was followed by Wilson Agar's single-barrelled 'coffee-mill' which was used by the Union forces in the Civil War and by the Gatling Gun, patented in 1862. This hand-cranked, six-barrelled machine gun could continue firing as long as the hand-operated crank was turned.

The cartridges were chambered, fed by gravity from its magazine, and fired one barrel at a time. Aside from mechanical problems, high rates of ammunition usage and expense, it suffered from not being considered central to battlefield dispositions and tactics. The Union's Chief of Ordnance, Brigadier-General James Ripley, was suspicious about new developments and opposed to the adoption of Agar's and Gatling's guns.

The same problem affected use of the French Mitrailleuse machine gun in the Franco-Prussian War of 1870–1. Although this crank-turned 37-barrelled gun could fire 44 rounds a minute, it was heavy, could jam and was insufficiently used in close support of troops. The failure of commanders to develop appropriate tactical uses for these early machine guns reduced the impact of this new technology.

Gatling guns, nevertheless, were used in 1873–4 in Garnet Wolseley's well-organized and successful punitive expedition against one of the more militarily powerful of all African people, the Asante (Ashanti) of West Africa. The guns were employed in the Zulu War of 1879, greatly strengthening British defensive positions. First used in combat there at the Battle of Nyezane River, they were put to use at Gingindlovu and Ulundi on the four corners of square defensive arrangements. Other hand-cranked machine guns included the Gardner, Lowell and Nordenfelt guns, the last of which was used by the British navy.

THE MAXIM GUN

Subsequently, machine guns became more powerful, although their use was still limited by their unreliability. In 1889, the British adopted a fully-automatic machine gun developed by the pioneering inventor Hiram Maxim after he was told, 'If you want to make your fortune, invent something to help these fool Europeans kill each other more quickly.' Patented by him in 1893 it could fire 600 rounds per minute, using the recoil to eject the empty cartridge case, replace it and fire. This was a single-barrel machine gun, unlike the Gatling, Mitrailleuse and Nordenfeldt. The Maxim was both reliable and readily transportable. It was water-cooled and fully-automatic. Early models of the gun readily broke and tended to be fouled by the black powder used as a propellant, so Maxim patented an improved gunpowder. Refinements in the manufacture of cartridges meanwhile reduced jamming.

The British employed the Maxim in Gambia in 1897, the Matabele War of 1893–4 in southern Africa, the Chitral campaign on the North-West Frontier of India in 1895, and at Omdurman in Sudan against the Mahdists in 1898. The British had 44 Maxim guns and 80 pieces of artillery at Omdurman including river gunboats. The poet, and later MP, Hilaire Belloc wrote 'Whatever happens we have got / the Maxim gun and they have not'; although, in practice, the breech-loader rifle shared in the carnage in Africa. While the British used machine guns there, the French did not, and the Germans were slow to do so. But in the 1890s, the Belgians used machine guns to help overcome opposition in the Congo. Aside from the Maxim, other machine guns in the closing years of the century included the Browning, which worked using barrel combustion gases, as did the Hotchkiss, while the Skoda's breech was blown back by propellant gases. The different operational methods provided manufacturers with a range of opportunities.

HEAVY CASUALTIES

The usage of the weapon reflected the breadth of Western imperialism and its merciless pace in this era. In 1901, Donald MacAlister, a member of the British field force sent against the Aros of southeastern Nigeria, noted 'We had the Maxim pouring into the bush this morning. There must be a great deal of dead.' Three years later, a British force advancing on Lhasa, the capital of Tibet, in order to thwart alleged Russian influence and dictate terms, opened fire at Guru on Tibetans who were unwilling to disarm. Due in large part to their two Maxim guns, four cannon and effective rifles, the British killed nearly 700 Tibetans without any losses of their own and pressed on to occupy Lhasa. In Kenya, in 1905, the British field force sent to suppress resistance among the nomadic Nandi used its ten Maxims to cause heavy casualties.

Machine guns

1718
James Puckle, a London lawyer, patents the 'Puckle gun' which fires nine rounds. It was never produced

1861
Richard Gatling takes out a patent for his Gatling gun. The hand-cranked weapon saw limited action in the American Civil War

1881
Hiram Maxim invents the Maxim gun – the first automatic machine gun which uses the recoil energy of the previous bullet to introduce the next

World War I
Machine guns based on the design of the Maxim gun are much feared, causing very heavy casualties. Lighter submachine guns are developed and machine guns are mounted on aircraft

1934
The German *Maschinengewehr 34* is the first general purpose machine gun

1947
The AK47 designed by Mikhail Kalashnikov becomes the weapon of choice during the Cold War famed for its rugged reliability and low cost

Specifications improved during the 20th century. Machine guns which, arguably, should have dominated the battlefield in the Boer War (1899–1902), did not. The Boers did not have ready access to advanced military technology and relied, instead, on their sharp-shooting. The British, who could have deployed machine guns in numbers, did not rely on them.

In contrast, in World War I, alongside artillery and rifles, they were to dominate the field. A technological leap forward was made between the Boer and First World Wars. In the Russo-Japanese War (1904–5), the fighting featured many elements that were also to be seen in World War I, including trench warfare with barbed wire and machine guns. The Japanese took the initiative, launched frontal assaults on entrenched forces strengthened by machine guns and quick-firing artillery, as at Port Arthur and Mukden in 1905, and prevailed despite horrific casualties.

In the Balkan Wars (1912–13), there was a general failure to note the degree to which machine guns and rapid-firing artillery might blunt massed infantry assaults. Instead, observers saw the wars as confirming their faith in massed infantry assaults. This lesson was taken in particular from the Bulgarian victories over the Turks in 1912, such as Kirkkilese and Lyule Burgas, which appeared to show the effectiveness of high morale and of infantry charging into the attack.

The Vickers-Maxim machine gun, adopted by the British army in 1912, fired 250 rounds per minute. Such guns equalled the fire of many riflemen. The machine gun became a metaphor of the application of industry to war. Although it was to be famous in World War I as a defensive weapon, before this the machine gun was seen as a useful tool for attack, especially by clearing ground of defenders (for this, the development of the sub-machine gun was vital). The rate of fire of this and other weapons ensured that supply needs for ammunition rose, and this led to pressure for a rapid victory. The weight of the guns and the fact that they were usually operated by a team – a loader as well as a gunner – reduced their effectiveness.

'A VERY HELL...'

Machine guns combined with artillery to make attacks in the war particularly deadly. Nearly nine million troops died in the war, including 27 per cent of all Frenchmen between the ages of 18 and 27. A British Master Sergeant in the Somme offensive in 1916 noted in his diary 'the whole place smells stale with the slaughter which has been going on for the past fourteen days. The place is a very Hell with the whistling and crashing of shells, bursting shrapnel and the rattle of machine guns. The woods we had taken had not yet been cleared and there were pockets of Germans with machine guns still holding out and doing some damage.' In the *Daily Express* of 5 July 1916, John Irvine informed readers of 'how terrible this machine gun fire has been... a hail of machine gun bullets which were simply terrific'. During the war, machine guns developed, new models including the German MG 08/15 and the American Browning Model 1917.

Machine guns were also used in air warfare. These were usually air-cooled, light machine guns, although the problems of firing through the arc of the propeller had to be overcome. The guns were used to attack other planes and also to strafe troops on the ground.

There was to be subsequent improvement in the deadly specification of the weapon, not least as a result of producing lighter machine guns that were more mobile, and also thanks to the production of armour-piercing ammunition. Key machine guns in World War II included among light machine guns the German MG-34 and MG-42 and the British Bren gun. The Vickers machine gun used by the British in World War II had a range of up to 4,000 yards and could fire 500 rounds a minute. In the Mauser MG-42, introduced in 1942, the Germans had a flexible, easy-to-use machine gun. This gave considerable strength to their defensive positions, and made it vital to suppress their fire before they were stormed. The MG-42 could fire 1,200 rounds per minute. It was operated by gas piston-action and was easy to make – from mass-produced pressed steel parts – and inexpensive. This was to be the basis of subsequent machine guns.

After World War II, there was an application of machine guns to the growing range of weapon platforms including helicopters. High rates of fire were achieved by a number of means including the use of chain guns which employed electrically driven roller chains which drove the weapons' bolts. Machine guns remain crucial to modern combat, although now the more easily carried sub-machine guns are commonly used.

Torpedoes

'Soon, soon this violent, terrifying thing would happen. I saw that the bubble-track of the torpedo had been discovered on the bridge of the steamer, as frightened arms pointed towards the water and the captain put his hands in front of his eyes and waited resignedly... then a frightful explosion followed, and we were all thrown against one another...'

U-BOAT ATTACK, APRIL 1916: ADOLPH KGE SPIEGEL

A WEAPON DESIGNED TO TERRIFY EVERY SAILOR was first devised by David Bushnell, an American who in the 1770s constructed the first operational turtle-shaped submarine. He described the armament as 'a torpedo', after the stinging crampfish of the genus *Torpedinidae*; it was, in fact, a mine. The problems of devising a system for propulsion held back work on torpedoes. In 1777, when using explosive charges, Bushnell had to rely on them floating. Subsequently, there was greater interest in underwater warfare, although Robert Fulton was held back in his experiments with torpedoes from 1807 onwards by his failure to devise an effective firing device.

The modern self-propelled torpedo originated with an Austrian invention of 1864 of a small vessel driven by compressed air and with an explosive charge at the head. When adopted by the Austrian navy in 1868, it was capable of a speed of 16 kilometres (10 miles) an hour. Austria had a coastline on the Adriatic and was a naval power there. The torpedo was manufactured by Robert Whitehead, a British expatriate. Britain (in 1872) and most European powers then bought the rights to manufacture it.

The first successful attack with a Whitehead torpedo occurred in January 1878, when the Russians sank the Turkish harbour guardship at Batumi on the Black Sea, with two torpedoes fired from launches; the British had launched an unsuccessful attack the previous May on a Peruvian warship that had seized British merchantmen. Torpedo boats developed in the late 1870s, with Thornycroft's *Lightning* (1876) providing the model. Two Chilean ships, part of the squadron blockading Callao, were sunk by Peruvian torpedoes in 1880.

However, an older form of torpedo was provided by those that were not self-propelled. In 1864, the first successful torpedo boat attack occurred in Albemarle Sound, North Carolina, during the American Civil War when, with a spar torpedo fitted to a steam launch, the Union sank the ironclad *Albemarle*. At Foochow in 1884, during the Sino-French War, two Chinese warships were sunk by French spar torpedoes.

THE ALTERNATIVE TO BATTLESHIPS?

There were also developments with mines. In the American Civil War, Confederate mines sank seven armoured Union ships, compared to one lost to shore batteries. In 1868, Russia designed a more effective mine, a glass-tube battery electrolyte detonation device.

Some commentators wondered if battleships had a future in the face of torpedoes. One opponent of battleships, Admiral Théophile Aube of the French navy, was very interested in the potential of torpedo boats, and claimed that they nullified the power of the British battleships. France rapidly responded to the British development of steam torpedo boats. It was not alone. By 1888, Germany had commissioned 72 torpedo boats and developed the manufacture of good torpedoes. This interest in an alternative to the battleship looked towards later German interest in the submarine.

In part due to the threat from the torpedo, the year 1887 was the only year between 1858 and 1922 in which no country laid down an armoured warship, but, in the 1890s, there was a renewed emphasis on battleships in naval doctrine. In part, this reflected growing awareness of means to thwart or limit torpedoes. These included torpedo nets and thick belt armour around ships' waterlines. Furthermore, electric searchlights and quick-firing medium-calibre guns were employed to provide a secondary armament for use against torpedo boats.

The platforms for torpedoes developed. The specialist torpedo destroyer, or 'all-purpose destroyer' as it later became, was seen as a fast-moving ship able to sink larger warships. They were used in combat for the first time in the Russo-Japanese War of 1904–5, with torpedoes employed for the successful surprise attack at the outset of the war on the Russian squadron in Port Arthur in the south of Manchuria. Submarines were also developed.

Improvements in the accuracy, range and speed of torpedoes followed. By 1914, they could travel 7,000 yards at 45 knots. As a result, submarines became more deadly. In October 1916, Admiral Jellicoe, the commander-in-chief of the British Home Fleet, wrote that the greater range and size of submarines, and their increased use of the torpedo, so that they did not need to come to the surface, meant that the submarine menace was getting worse.

WORLD WAR II

The same combination of improved torpedoes and better submarines was also at issue in World War II. For example, in 1943, the Germans introduced the T5 acoustic homing torpedo, at once sinking three escorts. The successful use of torpedoes required both effective weapons and appropriate tactics. Thus, in World War II, the Americans had initial problems with their torpedoes, as did the Germans and the Soviets. These affected operations, including those off the Philippines in the winter of 1941–2.

The Japanese, in contrast, were hindered by poor strategy. They had developed very fast, reliable and long-range oxygen-fuelled torpedoes in the 1930s, but they insisted on hoarding torpedoes for use against warships, rather than employing them against American logistical links. This was an aspect of the major failure in Japanese submarine warfare with their focus on warships reflecting an inability to appreciate the wider dimensions of warfare.

Torpedoes were used in the war not only by submarines, but also by aircraft, and by surface ships. Not only motor torpedo boats and destroyers, but also cruisers and battleships could have torpedo tubes. Destroyer torpedo attacks could be very effective, as when used by the Japanese against the Americans off Guadalcanal on 13 and 30 November 1942. Motor torpedo boats and other surface warships, however, were very vulnerable to air attack. Air support enabled the Japanese invading Hong Kong in 1941 to block interference with their supply routes from British motor torpedo boats.

Aircraft torpedo attacks proved disastrous for surface warships. The successful night attack by 21 torpedo planes on Italian battleships moored in Taranto on 11 November 1940 possibly inspired the Japanese attack on Pearl Harbor a year later; three battleships were badly damaged. On 28 March 1941, off Cape Matapan, thanks to torpedo aircraft as well as battleship firepower and ships' radar, the British sank three Italian cruisers and damaged a battleship. The German battleship *Bismarck*, which had broken out into the Atlantic, was crippled by a hit on the rudder by a British airborne torpedo attack (26 May 1941) before it was heavily damaged by British battleship fire. The *Bismarck* then fell victim to a cruiser-launched torpedo (27 May).

PEARL HARBOR

On 7 December 1941, the Japanese inflicted major damage on the American Pacific Fleet's major base, Pearl Harbor on the island of Oahu in the Hawaiian archipelago. Both torpedoes and dive bombers, from six Japanese carriers, were used in an attack in which two battleships were destroyed and three more damaged. This forced an important shift in American naval planning away from an emphasis on capital ships and, instead, towards carriers.

The attack also revealed grave deficiencies in Japanese (as well as American) planning, and also in the Japanese war machine. Only 45 per cent of Japanese naval air requirements had been met by the start of the war, and the last torpedoes employed in the attack were delivered only two days before the fleet sailed, while modification of planes to carry both torpedoes and heavy bombs was also last minute. Moreover, the Japanese bombers did not find the more crucial aircraft carriers, which were not in harbour. These carriers played a major role subsequently in checking, at the Battle of the Coral Sea, the Japanese advance into the South-West Pacific.

Tactically successful, the Japanese nevertheless had also failed to destroy Pearl Harbor itself. Because of the focus on destroying warships, there was no third-wave attack on the fuel and other harbour installations. Had the oil farms (stores) been destroyed, the Pacific Fleet would probably have had to fall back to the base at San Diego. Had the Japanese invaded the island of Oahu, the Americans would have had to do so, but the logistical task facing the Japanese in supporting such an invasion would have been formidable.

Furthermore, the course of the war was to reveal that the strategic concepts that underlay the Japanese plan had been gravely flawed. The Japanese underrated American economic strength and the resolve of the American people.

HUNTER-SUBMARINES

Torpedoes did not play a major role in conflict after 1945 because there was little naval warfare. However, torpedo effectiveness played a key role in the Cold War, as hunter-submarine forces were developed to attack hostile shipping and submarines. This effectiveness became particularly important as a result of the creation of submarine-based intercontinental missile capability, a capability that was best countered by submarine attacks.

The continued effectiveness of torpedoes was demonstrated in 1982 when HMS *Conqueror*, a British nuclear-powered submarine, controversially sank the Argentine cruiser *General Belgrano*, with the loss of over 300 Argentinians. This was crucial to the struggle for command of the sea, as it discouraged subsequent action by the Argentinian navy against the British task force preparing to recapture the Falkland Islands. Torpedoes remain central to submarine armament.

Submarines
World War I

'The submarine is, in fact, the true reply
to the submarine... that provided we are as
well equipped in the matter of submarines
as our neighbours, the introduction of this
new weapon, so far from being a
disadvantage to us, will strengthen our
position.'

H. O. ARNOLD FOSTER IN 1901
PARLIAMENTARY SECRETARY TO THE BRITISH ADMIRALTY

'**W**HAT IS IT THAT THE COMING OF THE SUBMARINE REALLY MEANS? It means that the whole foundation of our traditional naval strategy, which served us so well in the past, has been broken down!' (Memorandum in papers of British Admiral Jellicoe, 1912).

The first known description of a viable submarine was published by the English mathematician William Bourne in 1578, but a Yale graduate, David Bushnell (1740–1824), constructed the first operational machine. In 1774, he began to experiment with a submersible that would plant gunpowder beneath a ship, and the following year a prototype was ready for testing and a way had been found to detonate a charge underwater. The wooden submarine contained a tractor screw operated by hand and pedals, a surfacing screw, a drill for securing the explosive charge, fitted with a time fuse, to the hull of the target, a depth pressure gauge, a rudder with a control bar, bellows with tubes for providing ventilation, fixed lead ballast, detachable ballast for rapid surfacing and a sounding line.

Bushnell's *Turtle* could only attack ships at anchor, and was first employed, against HMS *Eagle* in New York harbour, on 6 September 1776 during the War of American Independence. Bushnell, however, encountered serious problems with navigating in the face of the currents and could not attach the charge, which went off harmlessly in the water.

Another American, Robert Fulton, produced a submarine in 1797, but found neither the French nor the British, then at war with each other, greatly interested in its acquisition, although this is less surprising given the high price he proposed for 'the destructive powers and easy practice of the engines'. His experiments for the French in 1800–1 included the testing of a system of compressed air in a portable container and the successful destruction of a vessel by an underwater explosion. During the Anglo-American War of 1812, Fulton also played a role in unsuccessful experiments with a submarine, mines and underwater guns.

CHARLESTON HARBOUR ATTACK

Effective submarines depended on a number of different technologies which came together in the late 19th century. In 1864, the modern self-propelled torpedo originated, with the invention of a submerged self-propelled torpedo driven by compressed air and armed with an explosive charge at the head. That year also saw the first effective attack by a submersible, mounted in Charleston Harbour, South Carolina during the American Civil War, when the *Hunley* sank the Union screw sloop *Housatonic*, although she herself sank soon afterward, probably as a consequence of the stresses created by the explosion. The first steam-powered submarine, the 30-ton *Resurgam*, was launched by George Garrett in 1879. Working with the Swedish arms manufacturer Thomas Nordenfelt, Garrett began work in 1882 on the *Nordenfelt I*, a 60-ton submarine, the first to be armed with self-propelled torpedoes. France completed a submarine powered by an electric battery in 1888.

Submarines were first employed to strategic ends in World War I. They had not featured prominently in naval operations over the previous decade, in, for example, the Russo-Japanese or Balkan Wars. Indeed their potential had been greatly underestimated by most commanders. Admiral Tirpitz, the head of the German navy, was a late convert to submarines. Britain, which had only launched its first submarine in 1901, had the largest number – 89 – at the outbreak of World War I.

There had been considerable speculation prior to the war about the likely impact of submarines, but scant experience on which to base discussion. In 1901, H.O. Arnold Forster, parliamentary secretary to the British Admiralty, was interested in how best to counter submarines, and saw other submarines as the best option:

The submarine is, in fact, the true reply to the submarine… That provided we are as well equipped in the matter of submarines as our neighbours, the introduction of this new weapon, so far from being a disadvantage to us, will strengthen our position. We have no desire to invade any other country: it is important that we ourselves should not be invaded. If the submarine proves as formidable as some authorities think is likely to be the case, the bombardment of our ports, and the landing of troops on our shores will become absolutely impossible.

The submarine was not in fact to be a key bar to invasion.

FOOD BLOCKADE

In World War I, German submarines swiftly affected the conduct of naval operations. The British Grand Fleet was obliged to withdraw from the North Sea and its base of Scapa Flow in the Orkneys in 1914 to new bases on the northwest coast of Scotland due to the threat of submarine attack: the fleet did not return to Scapa Flow until 1915, when its defences had been strengthened. Concern about submarines led Jellicoe to observe that year 'I am most absolutely adverse to moving the Battle Fleet without a full destroyer screen'. Submarines were seen by the Germans as a substitute for success in surface warship operations which had stalled in 1916 with the indecisive Battle

In the spring of 1917, British leaders, including Jellicoe, were pessimistic about the chances of success against the submarines, and the initial rate of Allied shipping losses was formidable. From February to April 1917, 1,945,240 tons of shipping were sunk with only nine submarines lost.

of Jutland, and also as a way to knock Britain out of the war. In 1915, the Germans tried unrestricted submarine warfare as part of a deliberate campaign to starve Britain into submission by blocking her food imports, but they stopped this on political grounds as likely to provoke American intervention.

In 1917, they returned to this policy. The Germans anticipated a rapid victory through the economic warfare of unrestricted attacks declared on 1 February, but the British did not sue for peace on 1 August, as it had been claimed they would. The Germans in practice had insufficient submarines to match their aspirations, in part because of problems with organizing and supplying construction, but largely because of a lack of commitment from within the navy to submarine warfare and a preference, instead, for surface warships. Although the Germans stepped up submarine production once war had begun, relatively few were ordered, and most were delivered behind schedule, due to poor organization and a concentration of industrial resources on the army.

Nevertheless, greater numbers ensured that more damage was done to British shipping than in 1915. In the spring of 1917, British leaders, including Jellicoe, were pessimistic about the chances of success against the submarines, and the initial rate of Allied shipping losses was formidable. From February to April 1917, 1,945,240 tons of shipping were sunk with only nine submarines lost.

The operational flaws of German warmaking rested on a strategic misjudgment of British vulnerability. During the war, the Germans sank 11.9 million tons of, mostly commercial, Allied shipping, at the cost of 199 submarines. However, aside from the benefits that stemmed from the British introduction of escorted convoys in May 1917 – in cutting shipping losses dramatically and in leading to an increase in the sinking of German submarines – the British were less vulnerable to blockade than anticipated. It proved possible to increase and reorganize food production in Britain, and to direct its distribution successfully. The German methods, but not capability, suggested total war, whereas the British institutional response, including the establishment of an effective Food Production Department, was far more a display of capability.

AMERICAN ENTRY

Politically the new policy outraged American public opinion. Americans became persuaded of the dangerous consequences of German strength and ambitions, and, by taking the high moral ground encouraged large-scale commitment. The USA entered the war on 6 April, and this added the world's strongest economy and third largest navy to the Allied war effort. American entry helped against the submarines for, from May 1917, American warships contributed to anti-submarine patrols in European waters. Brazil also suffered from the unrestricted submarine warfare and, in October 1917, it too declared war on Germany.

Anti-submarine capability improved greatly during the war. Mines sank more submarines than other weapons, and mine barrages limited the options for submarine warfare. Massive barrages that reflected industrial capacity and excellent organization were laid across the English Channel at Dover in late 1916, and across the far greater distance of the North Sea between the Orkneys and Norway from March 1918. Another barrage was laid in the Straits of Otranto in order to keep German and Austrian submarines in the Adriatic Sea and prevent them from breaking out into the Mediterranean. As an example of incremental development and of the application of scientific knowledge by the end of the war, magnetic mines had been developed, and were being laid by the British. Furthermore, aircraft and airships supported convoys in coastal waters, forcing submarines to remain submerged, where they were much slower.

Convoys reduced the targets for submarines and also ensured that when they found them they could be attacked by escorts. Convoys also benefited from the 'shoal' factor: when they found a convoy, submarines only had time to sink a limited number of ships. Only 393 of the 95,000 ships that were to be convoyed across the Atlantic were lost, including just three troop transports.

Although in the first four months of the unrestricted German submarine attacks in 1917, the British lost an average of 630,000 tons of shipping, the monthly tonnage fell below half a million in August 1917. Submarine operations declined much more in effectiveness towards the close of the war. Having sunk at least 268,000 tons of shipping each month from January to August 1918, only 288,000 tons were sunk in September and October combined.

Submarine operations had not swayed the course of the war on land: Allied land forces were able to maintain the offensive launched that summer, while large numbers of American troops continued to cross the Atlantic. The impact of the submarine was seen in the peace terms in 1919 under which the defeated Germans were banned from having any.

Radio

'... it was an epoch in history. I now felt for the first time absolutely certain that the day would come when mankind would be able to send messages without wires not only across the Atlantic but between the farthermost ends of the earth.'

GUGLIEMO MARCONI ON HEARING THE FIRST RADIO SIGNAL
ACROSS THE ATLANTIC, 12 DECEMBER 1901

THE ADVENT OF RADIO WAS AN EXCITING ADVANCE on existing methods of communication in military operations. In 1892, the first transmission occurred, and Guglielmo Marconi was soon able to transmit messages across steadily greater distances. In 1899, he transmitted across the English Channel and, in 1901, sent radio signals over 3,000 miles across the Atlantic Ocean. This was an important stage of development in communications, one that was not dependent on fixed links, as telegraphs were with their lines or undersea cables, both of which could be cut.

Although wireless telegraphy, which used Morse code, was applied successfully during the Russo-Japanese War of 1904–5, the British navy, which needed to control and co-ordinate widely separated units, was to be Marconi's best client. This was a key aspect in a major change in British naval strategy. During the late 19th century, this had been based on the deployment of much of the fleet on distant stations where squadrons, often substantial, were retained in order to support and further British imperial interests against the prospect of threats. Thus, there was a squadron in the Far East, one in the Indian Ocean, and so on.

COMMAND AND CONTROL REVOLUTION

In its place, thanks to radio, came a plan for a more flexible deployment. Many of the warships were to be kept in home waters, and they were to be sent to handle tasks rapidly in response to radio messages. This was a command and control revolution, and it was focused on the use of the new warships being built in this period, especially the Dreadnought-type battleships.

This strategic shift was also linked to a huge change in British strategy. From 1904, the focus of concern ceased to be France and, instead, became Germany, which was seen as an increasingly ambitious, powerful and unpredictable state. This encouraged an emphasis on home waters, as did the *entente* (easing of relations) with Russia in 1907. Whereas, in the Crimean War with Russia there had been, for example, concern about a possible Russian naval attack on Sydney, and in the 1880s on British India, the new strategic situation focused on preventing German warships from sailing from their bases in the North and Baltic Seas. This led the British to develop their naval bases in the North Sea.

More generally, radio networks were created prior to World War I. In 1912–14, the Germans built a network of radio stations in their colonies: at Duala, Windhoek, Dar-es-Salaam, Tsingtao, Yap, Apia, Rabaul and Nauru.

Alongside the development of wireless telegraphy, radio telephony was pioneered from 1904. The former was employed particularly for naval operations during World War I. As a result, the control of radio stations was important in operations outside Europe, and the Allies made major efforts to capture the German ones in 1914. The attack on German East Africa (now Tanzania) began with the shelling of the wireless tower at Dar-es-Salaam.

CONTINUOUS WAVE SYSTEM

In World War I, developments with radio made it easier to retain detailed operational control on land, at sea and in the air. Directional wireless equipment aided location and navigation, while radio transmissions changed from a spark message to a continuous wave system. The Allies made the most effective use of radio at sea, particularly to co-ordinate shipping. This was important not only in operations against German warships, but also in those against submarines and in convoy protection.

On land, the value of radio was not adequately grasped. Deficiencies in communications fed directly into command problems, particularly those of maintaining momentum in offensives. It was difficult not only to achieve breakthrough through the opposing trench line, but also then to sustain and develop it. This was a matter not only of the problems of advancing across terrain badly damaged by shellfire and maintaining the availability of shells for the all-crucial artillery, but also of providing adequate information to commanders about developments. This would have helped in enabling them to direct reserves to the correct place, and thus maintain the tempo of the advance. These issues were hard to overcome, although in 1918 the Allies proved successful on the Western Front. By then, the potential of radio was being demonstrated in land operations: it had come a long way since 1914.

During World War II, radio played a key role at the tactical and operational levels. For example, radio was used for carrier operations and anti-submarine patrols, as well as for directing airplanes, and for directing artillery fire, in the last role superseding field telephones.

Cost, working problems and a reluctance to adopt innovations, not least the size of early radios, affected their use. In the campaign against the Faqir of Ipi in Waziristan on the North-West Frontier of India in the late 1930s, radio was used by the British, but it was still in its infancy. As a result, in this campaign, only the largest bases and headquarters had reliable radio communications, so most signalling below brigade level was carried out using old-fashioned coloured flags, as well as the heliograph and dispatch riders.

Progress, nevertheless, was made during the 1930s, not least in using radios to control tank advances, a practice that was to become increasingly important in World War II. The British conducted manoeuvres with a tank brigade directed by radio telephony in 1931 and the Germans with a tank division in 1935. Radio was also increasingly seen as vital to combined operations, especially land–air co-ordination. The Germans, Americans and British made major advances, but these were not matched by the French, Italians, Japanese and Soviets. Instead, there was a reliance on human communications – by couriers – and on telephones, the former slow, and the latter easily interrupted.

Radio, however, still faced problems, not least that of the effects of climate, especially in the Tropics. This was far more important than in World War I because of the major role of Japan in the new war. The geographical range of the Pacific War ensured that issues of communication and co-ordination were particularly acute.

ENCRYPTION

There were also serious problems with security. Encryption took time and was not foolproof, as the Allied interception of German material amply demonstrated, while transmitting in clear was very risky. Transmission by ships and submarines permitted the fixing of their location, and was to play a major role in British anti-submarine operations.

Variations in the availability of such material proved very important. It was possible to plot U-boat positions from August 1941, but, in February 1942, a new, more complex code was introduced, and this was not broken until December 1942. This greatly helped in the defence of the Atlantic supply route.

During World War II, radio played a key role at the tactical and operational levels. For example, radio was used for carrier operations and anti-submarine patrols, as well as for directing airplanes and for directing artillery fire, in the last role superseding field telephones. Radio contributed greatly to the speeding up of tactical moves and to the co-ordination required for combined operations. Artillery fire, particularly that of the Americans, benefited from improved aiming and range that reflected not only better guns, but also radio communication with observers and meteorological and survey information. The Americans, with their high-frequency radios, were particularly adept at this. American forward air controllers in light aircraft using SCR-522 radios, were used to feed information to artillery fire direction centres.

NAVIGATION SYSTEMS

Radio navigation systems were particularly useful for aircraft, although they were also fitted to ships. These systems dealt with the problems of flying without being able to see location-fixing positions on the ground, in other words flying by night or over cloud or over large bodies of waters. As another instance of the frequency with which non-military advances were crucial in warfare, the use of radio navigation systems in World War II owed much to the 1930s development of long-distance civilian transport. As a result, at the outset of the war, there was a large-scale use among bombers of radio direction finders. These gave bearings on ground-based radio beacons, but were of limited accuracy at a distance from them. As a consequence, more sophisticated radio range systems were produced, linking receivers on planes to beams transmitted from ground beacons, with the planes instructed to drop their bombs when the beams intersected. These systems were developed by the Germans and were used by them with deadly effect in the bombing of Britain, but they were eventually countered by British jammers. The British lacked any counterpart, until 1942 when the British developed GEE, their own system of ground transmitters, pulses and airplane receivers.

Radio continued to be very important after World War II and it proved particularly significant both for combined operations and in helping to make command structures more flexible. This was a matter not simply of ready response to changing circumstances, and thus getting inside the opponent's decision-making process by moving faster; but also in granting low-level commanders a greater ability to act appropriately.

By the 1990s, real-time communications for individual units was extended to include material provided by satellites. Improvements in radio links increased the effectiveness of, for example, planes, although, under the crisis of conflict, there were problems with durability. These, for example, affected the radios used by the British army. Nevertheless, thanks to radios, co-ordination in the field has greatly improved over the last century.

Aircraft
World War I

MANNED HEAVIER-THAN-AIR FLIGHT, first officially achieved by the American Wright brothers in 1903, led the British press baron, Lord Northcliffe, to remark 'England is no longer an island'. Flight had had an earlier role in warfare with balloons, which were used by the French in the 1790s, but its capability was now transformed. Imaginative literature, such as that of H.G. Wells, had prepared commentators for the impact of powered, controlled flight, and it developed rapidly. In 1909, Louis Blériot made the first aeroplane flight across the English Channel, and a British report on changes in foreign forces during 1910 noted,

> *Great activity has been displayed in the development of aircraft during the year, particularly in France and Germany. The main feature in the movement has been the increased importance of the aeroplane, which in 1909 was considered to be of minor military value. This [importance] was due partly to the surprising success of the aeroplane reconnaissances at the French manoeuvres, and partly to the successive disasters of the Zeppelin dirigibles [gas-filled airships]… Aviation schools have been started in almost every country.*

DEFENCELESS TARGETS

Aviation rapidly became a matter for international competition and therefore anxiety. There was grave concern in Britain in 1909 and 1913 about the possibilities of an airship attack and the bombing of defenceless strategic targets and cities; although, in 1909, when the chief of the British Imperial General Staff sought views on the likely effectiveness of airships and planes, he met with a sceptical response from General Ian Hamilton, who was unimpressed about the possibilities of bombing. Hamilton wrote, 'the difficulty of carrying sufficient explosive, and of making a good shot, will probably result in a greater moral than material effect'. In 1911, Britain established an air battalion, and, in 1912, the Royal Flying Corps. General Ferdinand Foch, director of the French École Supérieure de la Guerre, argued in 1910 that air power would only be a peripheral adjunct to the conduct of war, but, despite this, the French Directorate of Military Aeronautics was created in 1914.

Aircraft were first employed in war during the Italian–Turkish war in 1911 and the Balkan Wars of 1912–13. Grenades were dropped from the air on a Turkish army camp on 23 October 1911, and Turkish-held Edirne (Adrianople), besieged by the Bulgarians in 1913, was the first town on which bombs were dropped from an aeroplane. By 1914, the European powers had a total of over 1,000 aeroplanes in their armed forces: Germany had 245, Russia 244, France 141, Britain 113 and Austria about 55.

Aircraft were first used on a large scale in World War I, in which they played an important role, not only in fighting other aircraft, but also in influencing combat on the ground (and at sea), with reconnaissance aircraft proving especially valuable, particularly in helping direct artillery fire. Aircraft were used for spotting and reconnaissance in 1914, which was how the fighter evolved: as an armed reconnaissance plane protector, followed by armed aircraft for shooting down spotters. The value of aircraft reconnaissance was quickly appreciated, not least because in 1914 aircraft provided intelligence on the moves of armies in the opening campaign on the Western Front, particularly the change of direction of the German advance near Paris. The following year, General Callwell, critical of the initial Allied plan for the attack on Gallipoli as a purely naval operation, remarked 'As a land gunner I have no belief in long range firing except when there are aeroplanes to mark the effect'. The Turkish columns advancing on the Suez Canal in 1915 were spotted by British planes.

'Seeing over the hill' altered the parameters of conflict, but, despite being able to strafe troops and tanks, aircraft were not yet a tactically decisive or operationally effective tool. Their role had been grasped, but execution was limited.

AERIAL COMBAT

The ability of aeroplanes to act in aerial combat, nevertheless, was also enhanced. Increases in aircraft speed, manoeuvrability and the maximum height that a plane could fly, made it easier to attack other planes. Engine power increased and size fell, while the rate of climb of aircraft increased. Synchronizing gear, developed by Anthony Fokker, and modelled on a French aircraft shot down by the Germans, was used by the Germans from 1915 and copied by the British. It enabled aeroplanes to fire forwards without damaging their propellers. The Fokker Eindecker aircraft, which the Germans deployed from mid-1915, gave them a distinct advantage, and enabled them to seek the aerial advantage over Verdun in the key battle in 1916. Harold Wyllie, a squadron commander in the British Royal Flying Corps, wrote in 1916, 'sending out F.E.'s [F.E. 26s] in formation with Martinsydes for protection is murder and nothing else'.

Fighter aircraft of World War I

1914 The French Nieuport 10 is used by Allied forces at the outbreak of war for observation

1915 April The Fokker Eindecker (Monoplane) enters service with the German air force, its synchonizing gearing enables a machine gun to fire through the propeller

1915 June The de Havilland DH2 arrives at the front equipped with a 'pusher' rear-facing propeller

1916 Jan The Nieuport 11 enters service, designed to compete with the Fokker Eindecker

1916 Aug The new Albatross D-I regains the air initiative for Germany

1916 Oct The Sopwith Pup is introduced by the Royal Flying Corps

1917 March The SE5a enters service helping to wrest air superiority from the Germans

1917 May Captain Albert Ball VC, credited with 44 victories, is killed

1917 June The twin-gunned Sopwith Camel replaces the Pup. Over 5,000 are eventually produced

1917 June Manfred von Richthofen becomes commander of Jagdgeschwader 1, a formation which becomes known as the Flying Circus

1917 Sept The first highly manoeuvrable Fokker Triplanes are introduced

1918 April Manfred von Richthofen is killed by a single bullet fired from the ground

The eventually successful French attempt to contest the German advantage reflected their deployment of large groups of aircraft and the fact that they now also had planes with synchronized forward-firing machine guns, all of which allowed them to drive off German reconnaissance airplanes. In turn, in the winter of 1916–17, the Germans gained the advantage, thanks in part to their Albatross D-1, only to lose it from mid-1917 as more and better Allied planes arrived. Despite this, the Germans did not lose in the air as they were to do in World War II, and this largely indicated the relatively more limited capability of World War I aircraft.

There were also developments in tactics during the war. Aeroplanes flew in groups and formation tactics developed. Aeroplanes also became the dominant aerial weapon: their ability to destroy balloons and airships with incendiary bullets spelled their doom.

The British Royal Naval Air Service conducted the first effective bombing raids of the war when planes carrying 20-pound bombs flew from Antwerp to strike Zeppelin shells at Düsseldorf and destroyed an airship. The Germans launched bomber attacks on London in 1917 because they believed, possibly due to reports by Dutch intelligence, that the British were on the edge of rebellion. The attacks were intended not so much to serve attritional goals, but rather to be a decisive war-winning tool destroying the population's morale.

The use of bombers, the German Gotha, reflected the rapid improvement of capability during the war, as science and technology were applied in the light of experience. The Gotha Mark Four could fly for six hours, had an effective range of 520 miles, could carry 1,100 pounds (or 500 kg) of bombs, and could fly at an altitude of 21,000 feet (four miles or 6,400 metres), which made interception difficult. The crews were supplied with oxygen and with electric power to heat the flying suits. The first (and deadliest) raid on London, a daylight one on 13 June 1917, in which 14 planes killed 162 people and injured 432, not least as a result of a direct hit on a school that killed 16 children, led to a public outcry and was met – in the rapid action-reaction cycle that characterized advances during the war – by the speedy development of a defensive system involving high-altitude fighters based on airfields linked by telephones to observers. This led to heavy casualties among the Gothas and to the abandonment of daylight raids. More seriously, the rationale of the campaign was misplaced because, far from hitting British morale, the bombing led to a hostile popular response. This remained the case even in the winter of 1917–18, when the Germans unleashed four-engine Zeppelin-Staaken R-series bombers, able to fly for ten hours and to drop 4,400 pounds (or 2,000 kg) of bombs.

By the close of the war, the extent and role of air power had dramatically expanded. By 1918, the British had 22,000 aeroplanes, and, that September, a combined Franco-American-British force of 1,500 was launched against the Germans in the Saint-Mihiel Salient, the largest deployment so far. In 1917, German aeroplanes destroyed French tanks in Champagne. Supply links came under attack from the air, inhibiting German and Austrian advances in 1918.

MASS PRODUCTION

Aeroplane production had risen swiftly. In 1914, the British Royal Aircraft Factory at Farnborough could produce only two air-frames per month, but their artisanal methods were swiftly swept aside by mass production. Air power also exemplified the growing role of scientific research in military capability: wind tunnels were constructed for the purpose of research. Strutless wings and aeroplanes made entirely from metal were developed.

The alarm raised in sections of British society by German air attacks encouraged post-war theorists to emphasize the potential of air power. During the war itself, however, the consequences of strategic bombing – either to disrupt industrial life or to cause civilian casualties – was limited. For 1919, the British had planned long-range bombing raids on German cities, including Berlin, with large Handley Page VI-500 bombers, but the war ended before their likely impact could be assessed, although one of the planes successfully flew the Atlantic in 1919.

Aircraft were used extensively after the war for military tasks. The British air force bombed Jalalabad and Kabul during the Third Afghan War in 1919, tribesmen in Central Iraq in 1920, and Wahabi tribesmen who threatened Iraq and Kuwait in 1928. In Somaliland, the Dervish stronghold at Taleh was bombed in 1920. Air power had become a vital ingredient for imperial control. However, in 1922, the General Staff of the British Forces in Iraq observed, in a military report on part of Mesopotamia, 'Aeroplanes by themselves are unable to compel the surrender or defeat of hostile tribes', a lesson that repeatedly needs to be re-learned.

Tanks
World War I

'We heard strange throbbing noises, and
lumbering slowly towards us came three huge
mechanical monsters such as we had never
seen before... big metal things they were,
with two sets of caterpillar wheels that went
right round the body.'

BERT CHENEY ON THE FIRST TANKS ON 15 SEPTEMBER 1916

REALLY A FEARSOME SIGHT... The road was on a slope of the hill, and the tanks just crawled up the slope, up the right bank nose in air, down with a bump into the road and across it – almost perpendicularly up the left bank, and down with a bump behind it and so onward up the hill without a moment's pause or hesitation.'

Edward Heron-Allen's account of tanks crossing a road on 16 October 1918 ably described the subordination of terrain by the new weapon. Prior to World War I, there had been interest in armoured vehicles. In 1909, Colonel Frederick Trench, the British military attaché in Berlin, reported that the Germans were proposing to subsidize power traction vehicles 'of a type suitable for military use'.

Such interest was spurred on during the war as both sides sought comparative advantage and also in response to the problems posed by trench warfare. In January 1915, Arthur Balfour, a member of the British Committee for Imperial Defence, argued, 'that the notion of driving the Germans back from the west of Belgium to the Rhine by successfully assaulting one line of trenches after another seems a very hopeless affair, and unless some means can be found for breaking their lines at some critical point, and threatening their communications, I am unable to see how the deadlock in the West is to be brought to any rapid or satisfactory conclusion.' The previous month, the Committee's secretary, Maurice Hankey, had suggested,

> Numbers of large, heavy rollers, themselves bullet proof, propelled from behind by motor engines, geared very low, the driving wheels fitted with 'caterpillar' driving gear to grip the ground, the driver's seat armoured, and with a Maxim gun fitted. The object of this device would be to roll down the barbed wire by sheer weight, to give some cover to men creeping up behind, and to support the advance with machine gun fire.

SHOCK TACTICS

Tanks were invented independently by the British and French in 1915, and were first used in combat, at the Battle of the Somme, on 15 September 1916, and by the French the following April. The British use of them *en masse* at Cambrai in November 1917 was certainly a shock to the Germans. On 8 August 1918, no fewer than 430 tanks broke through the German lines near Amiens. That November, the French planned to deploy 600 tanks to support an advance into Lorraine, and by 1918 they had 3,000 tanks.

The Germans deployed tanks in 1918, but did so in far smaller numbers, and to less effect than the British: these tanks, some captured from the British, did not influence the outcome of the German spring 1918 offensives. German industry was unable to manufacture tanks in sufficient quantities.

Tanks could be hit by rifle bullets and machine guns without suffering damage. They could also smash through barbed wire and cross trenches. Tanks seemed to overcome one of the major problems with the offensives against trenches that had up to then proved so unsuccessful and costly in manpower during World War I: the separation of firepower from advancing troops, and the consequent lack of flexibility. Instead, by carrying guns (cannon) or machine guns, tanks made it possible for advancing units to confront both positions hitherto unsuppressed by artillery fire and counter-attacks. Indeed tanks offered precise tactical fire to exploit the consequences of the massed operational bombardments that preceded attacks.

Tanks offered mobility, not only in breaking open a static battlefield but also in subsequent operations. A memorandum of June 1918 from the British Tank Corps Headquarters claimed,

> Trench warfare has given way to field and semi-open fighting ... the more the mobility of tanks is increased, the greater must be the elasticity of the co-operation between them and the other arms. The chief power of the tank, both material and moral, lies in its mobility, i.e. its pace, circuit, handiness, and obstacle-crossing power.

Now the tank commander had to make sure he was not too far in advance of the infantry,

> whilst formerly he merely led the infantry on to their objective protecting them, as best he could, now he must manoeuvre his tank in advance of them zig-zagging from one position to another, over-running machine guns, stampeding away and destroying the enemy's riflemen and all the time never losing touch with the infantry he is protecting. This increased power of manoeuvre of the Mark V Tank demands an increased power of manoeuvre on the part of the infantry.

The British light infantry mortar in practice was more effective, more reliable and more capable of providing flexible infantry support, than the tank, which was under-powered, under-gunned, under-armoured and unreliable. Moreover, it was difficult for the crew to communicate with each other, let alone with anyone outside the tank...

The value of tanks and their likely future consequences attracted much attention from commentators. Commanders had to decide how best to employ tanks, and how to combine them with infantry and artillery. The variety and development of new tank types made this a dynamic issue.

UNDER-POWERED, UNDER-GUNNED, UNDER-ARMOURED

Yet, as a reminder of the need to understand their limitations as well as potential, it is necessary to note how far the value of tanks was lessened by their problems, which included durability, but also firepower and speed. The British light infantry mortar in practice was more effective, more reliable and more capable of providing flexible infantry support, than the tank, which was under-powered, under-gunned, under-armoured and unreliable. Moreover, it was difficult for the crew to communicate with each other, let alone with anyone outside the tank, and this made it harder to get a tank to engage a target of opportunity.

The value of tanks was also affected by the difficulty of providing sufficient numbers of them, which reflected their late arrival in wartime resource allocation and production systems. Moreover, the British suffered from a failure to produce enough spare parts.

The ability to devise anti-tank tactics was also significant. German anti-tank measures were quite effective. Wherever tanks met real resistance, they did not do nearly as well as anticipated. The use of artillery against tanks was particularly important in this respect, and reflected the extent to which the incremental nature of improvements in artillery was a matter of tactics as well as technology and numbers.

To operate most effectively, tanks needed to support, and to be supported by, advancing artillery and infantry. This was a lesson that had to be learned repeatedly during the following century in the face of pressure from enthusiasts for tanks alone. For example, the British Eighth Army encountered serious problems in conflict with the German Afrika Corps in North Africa in 1941 because British armoured advances lacked sufficient support, and this strategy had to be altered before success could be achieved at El Alamein in 1942.

UNFIT FOR SERVICE

During World War I, British successes at Cambrai (1917) and Amiens (1918) provide misleading examples of the usefulness of tanks because they did not meet effective organized resistance, and most of the tanks engaged in these battles subsequently broke down or were otherwise immobilized within a few days. Many tanks broke down even before reaching the assault point and, in battle, tanks rapidly became unfit for service, understandably so given their technical problems. For these reasons, there was a reaction in British circles against their use after August 1918. The French tanks contributed little to the unsuccessful Nivelle offensive of 1917.

Had tank production been at a greater level, then tanks might have made a greater contribution in 1918, but the idea that massed tanks would have made a significant difference to Allied capability had the war continued into 1919 is contentious. Assuming that the tank could have been mass-produced in order to manufacture the huge numbers required, which was not previously the case, the same basic problems of unreliability – slow speed, vulnerability to anti-tank measures and field guns, under-gunning, poor inter-communication capabilities, and poor obstacle-crossing capability – would have still remained.

There is little to suggest that tanks would have performed well. If the British tanks of the 1920s are considered as an extension of the line of development from World War I, it is difficult to see how they would have been decisive. Indeed, in 1928, General Montgomery-Massingberd, head of Britain's Southern Command, noted of 'The latest Carden Lloyd tanks … a great advance… Our trial machine did 49 miles an hour, which for a track machine seems almost undreamable.' In addition, this approach ignores the anti-tank technologies that would have been developed by the Germans. Indeed, the chances are that anti-tank guns would have been superior to the tanks.

Nevertheless, it was widely believed that tanks were the weapon of the future and their use spread. The British trained a tank force in 1919 for General Denikin's anti-communist army in the Russian Civil War, and a tank was used to help thwart a coup in Ethiopia in 1928. However, given the limited prospect of a great power war in the 1920s, not least because Germany had been disarmed, tanks seemed most pertinent for imperial roles outside Europe. For example, Lieutenant-General Sir Philip Chetwode of the British army argued in 1921 that tank specifications and tactics ought to focus on colonial commitments, rather than the possibility of conflict with other regular forces. He pressed for tanks to be armed with a machine gun, not a heavier gun, and for training in the use of tanks against opponents equipped with artillery and machine guns, but not tanks. This advice reflected the flexibility of the developing weapon.

Anti-Tank and Anti-Aircraft Guns

'A tank had churned its way slowly round
behind Springfield and opened fire; a
moment later I looked and nothing remained
of it but a crumpled heap of iron; it had been
hit by a large shell.'

THE BATTLE OF LANGEMARK, 27 AUGUST 1917, EDWIN CAMPION VAUGHAN

ANTI-WEAPONRY MOVES IN CONCERT WITH WEAPONRY and the tank was no exception. This anti-weaponry tends to receive insufficient attention, which is a mistake as the anti-weaponry helps define the possibilities presented by existing and new weapons. The responses to the tank provide a good example. In World War I, a range of weapons was used against tanks including other tanks, mines, artillery pieces and machine guns. Armour-piercing bullets fired by the last, and low velocity shells from artillery pieces were of particular importance.

In the inter-war period, there was a development of high velocity guns that used solid shot in order to penetrate tank armour by kinetic energy. These guns were mounted in tanks and in carriages that were towed in the field. In 1934, the British Committee of Imperial Defence pressed for anti-tank guns for the infantry.

The availability or not of opposing anti-tank guns was crucial to the relative success of German offensives in the early stages of World War II. In 1939, the Germans benefited in their attack on Poland, because the Polish army was weak in anti-tank guns and training. However, as an important indication of the limitations of armour, a tank advance into Warsaw on 9 September was stopped in street fighting by Polish anti-tank guns and artillery. In their offensive on the Western Front in 1940, the Germans again succeeded because the British were short of anti-tank and anti-aircraft guns. Indeed, the British lacked an effective anti-tank gun.

MOLTEN GAS

Improvements in tank specifications during World War II created problems for anti-tank weaponry, with thicker armour leading to more powerful weapons. This was particularly the case with stronger Soviet and German tanks. In response, anti-tank guns developed, leading to guns with larger calibres (e.g. 105 mm German guns, not 88 mm), longer barrels and better projectiles. This entailed alternatives to solid armour piercing shot which was limited against hardened armour. Hollow charge chemical energy projectiles, which threw a jet of molten gas against armour, were one response.

The Americans first used the Bazooka anti-tank rocket in 1942, but failed to upgrade it as tanks got heavier. The Germans, however, developed the design into the more powerful Panzerschreck rocket grenade. They also developed the hand-held Panzerfaust rocket launcher. Other weapons against tanks included those mounted on aircraft, both guns and rockets, which were significant in the battle for Normandy in 1944. Tanks themselves were crucial weapons against tanks, but anti-tank guns were also important. They were less vulnerable and expensive than tanks. Motorized (self-propelled) tank destroyers also had a major impact: effective German versions were matched by American tank-destroyers armed with 76 mm and 90 mm guns.

> The use of barrages became more valuable as anti-aircraft guns became more powerful, shells became more effective, and the proximity fuse – a radio device which detonated projectiles within lethal range of their targets – was introduced in June 1944. It was used against the German V-1 rocket with considerable success.

INTIMATE CO-OPERATION

The effectiveness of anti-tank weaponry ensured that mixed formations were more effective than those that focused solely on tanks. In February 1945, the British Field Marshal Montgomery argued that close co-operation with infantry was needed in order to overcome anti-tank guns, 'I cannot emphasize too strongly that victory in battle depends not on armoured action alone, but on the intimate co-operation of all arms; the tank by itself can achieve little.'

Indeed commanders of armoured units urged their officers to wait for support rather than charging in. This was a sensible response to the German skill in defensive warfare, particularly the careful siting of anti-tank guns to destroy advancing tanks. In July 1944, Sir Richard O'Connor, the commander of the British 8th Corps in Normandy, instructed the commander of an armoured division to

> go cautiously with your armour, making sure that any areas from which you could be
> shot up by Panthers and 88s are engaged. Remember what you are doing is not a rush
> to Paris – it is the capture of a wood by combined armour and infantry.

Anti-aircraft guns had developed in World War I when rapid progress was made under the pressure of air attack. In 1918, the German Air Service's anti-aircraft guns shot down 748 Allied aircraft. Aside from the guns, there were specialized spotting and communication troops, as well as relevant training, manuals and firing tables.

Anti-aircraft guns became more effective during World War II as a result of the employment of radar. This permitted the projection of the height and course of hostile aircraft, and the estimate of future positions made it possible to plan effective interception by anti-aircraft fire. Radar-directed searchlights contributed to this defence.

SHELL BARRAGES

Another approach was to fire barrages of shells to provide a barrier to the aircraft, essentially using anew the techniques of World War I. The use of barrages became more valuable as anti-aircraft guns became more powerful, shells became more effective, and the proximity fuse – a radio device which detonated projectiles within lethal range of their targets – was introduced in June 1944. It was used against the German V-1 rocket with considerable success. Anti-aircraft guns indeed destroyed more of these rockets than the aircraft, which are usually regarded as the best interceptors.

The impact of air attacks on warships ensured that far more attention than before was given to anti-aircraft capacity at sea, although this was increasingly seen as most effective if in concert with fighter cover. In January 1942, Admiral Sir Geoffrey Layton, British Commander-in-Chief Ceylon, complained about the absence there of 'fast patrol craft with good AA [anti-aircraft] armament'. This became even more important in 1945 in response to Japanese *kamikaze* (suicide) air attacks on Allied warships. In response, the Allies dramatically increased the number of anti-aircraft guns on their ships, which meant that they required larger crews.

ARAB-ISRAELI WAR

After World War II, guns were progressively supplemented by missiles which proved a winner in combating tanks and aircraft. In the Arab-Israeli War of 1973, Israel's American aeroplanes and tanks proved vulnerable to Soviet ground-to-air (SAM-6) and Soviet Sagger ground-to-ground missiles, although the Israelis argued that the best anti-tank weapon was another tank: Israeli tank losses were 75 per cent to other tanks and only 25 per cent to missiles. In contrast, air losses were 5 per cent to other aircraft, 40 per cent to conventional anti-aircraft guns and 55 per cent to missiles.

SAM missiles had already inflicted heavy losses on American aircraft bombing North Vietnam in 1965–8. The use of electronic jamming in order to limit attacks by missiles and radar-controlled guns had considerable success for the USA, but the North Vietnamese learned, in part, to counter this by aiming at the jamming signals.

In Africa, SAM-7 missiles supplied by the Soviet Union helped weaken the Portuguese in their resistance to anti-colonial guerrilla movements in Angola, Guinea-Bissau and Mozambique in the early 1970s. These missiles countered the Portuguese use of tactical air support and helicopters. In Guinea-Bissau, SAM-7 missiles, available from 1973, not only challenged Portuguese air superiority but also contributed powerfully to the sense that the Portuguese had lost the initiative. In Mozambique, these missiles, available from 1974, also shifted the balance of military advantage.

In the Iran–Iraq war of 1980–8, the Iranians benefited from the use of missiles against Iraqi armour. Helicopters fired heat-seeking missiles against tanks, while SAM-7 surface-to-air missiles were also employed.

THE FALKLANDS WAR

Missiles proved less successful in the Falklands War of 1982, in which Britain regained islands in the South Atlantic that had been invaded by the Argentinians. Argentine air-launched missiles and bombs led to the loss of six British ships and the damage of another 11. These losses showed that modern anti-aircraft missile systems, in this case Sea Darts and Sea Wolfs, were not necessarily a match for manned aircraft, and revealed a lack of adequate preparedness on the part of the British navy, which had to rely on missile systems not previously tested in war.

Another failure of anti-aircraft defences occurred in 1991, when a major American-led air offensive was launched on Iraq in response to its failure to withdraw from Kuwait. This offensive worked because of the rapid success in overcoming the sophisticated Iraqi anti-aircraft system. Saddam Hussein had used French and Soviet technology to produce an integrated system in which computers linked radars and missiles. The destruction of the air-defence system on the first night was a triumph not only for weaponry but also for planning. The Coalition forces made full use of the opportunities presented by the weapons, while also out-thinking the Iraqis, for example by getting them to bring their radars to full power, and so exposing them to attack.

In Lebanon in 2006, in contrast, Hizbollah mines and its Soviet anti-tank missiles challenged the Israeli armour, the heavy Merkava tank which had hitherto provided a key capability advantage. The sophistication of anti-tank missiles has improved. There were improvements in wire-guided missiles, from the Soviet Sagger model introduced in 1963, to the more accurate American Two, introduced in 1970, and the Soviet Metis-M, introduced in 1978, as well as the development of laser-guided anti-tank missiles, such as the Russian Kornet-E, introduced in 1994. Thanks to such missiles, the tank increasingly became obsolescent in close combat, particularly in urban areas, in part because of the cost in resources and manpower entailed in losing tanks. Anti-tank and anti-aircraft weaponry continue to indicate the effectiveness and importance of 'anti' weapons.

Aircraft Carriers

'Looking about, I was horrified at the
ction that had been wrought in a matter
of seconds. There was a huge hole in the
deck just behind the amidship elevator...
Deck plates reeled upward in grotesque
igurations. Planes stood tail up, belching

AIRCRAFT CARRIERS TRANSFORMED NAVAL WARFARE and proved particularly effective in the Pacific in World War II. The transfer of new aerial capacity to naval warfare had been rapid. Aircraft were used in World War I for reconnaissance, for attacks against shipping and for patrols against submarines. These aircraft were land-based or seaplanes, or were based on particular warships. As yet, aircraft were not able to make a major contribution to anti-submarine operations.

Britain led in naval air power, with, by the close of the war, the only ship able to launch and land wheeled aircraft (as opposed to seaplane tenders) and 2,949 aircraft in the Royal Naval Air Service. Dedicated aircraft carriers, whose sole purpose was to act as a floating air base, had yet to come into service but two were under construction. In July 1918, Britain conducted the first raid by land planes flown off an improvised aircraft carrier and, in September, HMS *Argus*, a carrier capable of carrying 20 planes with a flush deck unobstructed by superstructure and funnels – the first clear-deck carrier – was commissioned by the British, although she did not undergo sea trials until October 1918. Britain also staged raids with carriers when she intervened against the communists in the Russian Civil War after the end of World War I.

NAVAL AIR POWER

There was no experience with conflict between aircraft carriers, but there was considerable confidence in their potential. In 1919, Admiral Jellicoe, who visited Australasia, pressed for a British Far East Fleet, to include four carriers as well as eight battleships, in order to deter Japan, while, in 1920, Rear Admiral Sir Reginald Hall MP argued in the *Times* that, thanks to aircraft and submarines, the days of the battleship were over. This, however, was of scant interest to the British Admiralty, which remained wedded to the battleship.

Nevertheless, despite financial stringency, there was an important commitment to naval air power. The number of planes in the Fleet Air Arm rose to 144 by the end of the 1920s, while new arrester gears were fitted which help slow the planes down and, in addition, they could be reset automatically. These were in use by 1933 to help in the difficult task of landing on a carrier. Four carriers were commissioned by Britain in 1923–30 converted from a battleship and three battlecruisers. British carriers were sent far afield in this period. HMS *Argus* was stationed near the Dardanelles during the Chanak crisis between Britain and Turkey in 1922. There was also a carrier on the China Station in the late 1920s (first HMS *Hermes* and then HMS *Argus*), and another, HMS *Furious*, took place in the major naval exercises in the late 1920s.

In the 1930s, however, the lead was increasingly taken by the USA and Japan, in part because they would be the key powers in any struggle for control of the Pacific. Indeed, War Plan Orange, the plan for war with Japan, was a vital element in American strategic planning.

Air power had been restricted in the 1910s and 1920s by the difficulty of operating planes in bad weather and the dark, by their limited load capacity and range, and by mechanical unreliability, but improvements were made, especially in the 1930s.

Whereas naval air power in Britain lacked a separate institutional framework, because the Fleet Air Arm was placed under the Royal Air Force (RAF) between 1918 and 1937, and the RAF, primarily concerned with land-based planes had little time for their naval counterparts; in the USA there was a very different situation thanks to the Bureau of

Aeronautics of the American navy. This stimulated the development of effective air-sea doctrine, operational policies and tactics. Aside from advances with carriers, there were also marked improvements in aircraft. Because of this the Americans benefited from the development of dive-bombing tactics in the 1920s and, subsequently, of dive-bombers. These proved more effective than torpedo-bombers which were vulnerable to defensive power.

JAPANESE THREAT

No naval power was as acute a threat to Britain in the 1920s as Japan was to the USA. Furthermore, in the 1930s, the need for British deep-sea air capacity provided by carriers in any war with Germany appeared lessened by the vulnerability of German naval power to land-based air attacks. The lesson of World War I appeared to be that the Germans could be bottled up in the North Sea. In 1940, however, they transformed the situation by seizing first Norway and then France, and then basing powerful warships and submarines in both.

Despite this, the British, who were concerned also about Italian and Japanese intentions, developed their carrier capacity in the late 1930s, giving Britain an important added dimension to her naval superiority over other European powers: France had only one carrier. Four 23,000-ton carriers, the Illustrious class, each able to make over 30 knots and having a 3-inch armoured flight deck, were laid down in 1937, following the 22,000-ton HMS *Ark Royal* laid down in 1935. The 23,450-ton HMS *Implacable* was laid down in February 1939, although it was not commissioned until August 1944.

The damage caused by German air attacks on the British fleet in the early stages of World War II demonstrated the value of air support. The British doctrine of reliance on anti-aircraft fire was revealed as inadequate, and Admiral Sir Dudley Pound, the First Sea Lord, remarked, 'The lesson we have learnt here is that it is essential to have fighter protection over the Fleet whenever they are within reach of the enemy bombers'.

STEEP LEARNING CURVE

The Germans, like the Italians and the Soviets, had no carriers, and conflict between carriers was therefore restricted to the American–Japanese war in the Pacific (1941–5), and the smaller Anglo-Japanese equivalent in the Indian Ocean. Limited pre-war experience meant that there was a steep learning curve for both sides. A variety of factors, aside from the key American ability to build more carriers, was important to American victory. The Japanese were also hit by flawed preparation and planning. For example, in the pivotal Midway campaign in 1942, the Japanese underestimated American strength, while their deployment in pursuit of an overly complex plan and their tactical judgment were both very poor. Admiral Yamamoto also exaggerated the role of battleships in any battle. American tactical flexibility was also fundamental, as the ability to locate opposing ships and planes, and to respond to both, proved crucial at Midway, as elsewhere.

The sinking of four heavy Japanese carriers at Midway on 4 June 1942 shifted the naval balance in the Pacific to the Americans. Both the initiative and the arithmetic of carrier power moved against the Japanese. The loss of pilots was particularly serious as the Japanese had stressed the value of training and had produced an élite force of aviators. Once lost, they proved difficult to replace.

During the war, the Americans developed self-sufficient carrier task-groups (supported by at-sea logistics groups), which did not depend on a string of bases. This allowed their rapid advance across an unprecedented distance, with a lot of island-hopping that destroyed

The sinking of four heavy Japanese carriers at Midway on 4 June 1942 shifted the naval balance in the Pacific to the Americans. Both the initiative and the arithmetic of carrier power moved against the Japanese. The loss of pilots was particularly serious as the Japanese had stressed the value of training and had produced an élite force of aviators. Once lost, they proved difficult to replace.

any hope that the Japanese might retain a defensive perimeter in the Pacific. Thus, in the Battle of the Philippine Sea in June 1944, an American fleet with 15 carriers and 902 aircraft devastated its Japanese opponent, which had nine carriers and 450 aircraft, and this enabled the Americans to overrun the Marianas Islands, a decisive advance of Allied power into the western Pacific. The Japanese carriers were protected by a screen of Zero fighters, but, as a clear sign of growing Japanese weakness in the air, this was too weak to resist the fighters that supported the American bombers. Once again, the loss of Japanese pilots proved a crippling blow.

Carriers also played a major role in the Allied struggle against German submarines, especially in the Battle of the Atlantic, providing valuable air-cover. The first escort aircraft carrier entered service in late 1941.

DECLINE OF THE BATTLESHIP

After World War II, the decline of the battleship ensured that carriers were the big ships. The USA dominated carrier capability, and planned the use of its carrier strength for a variety of purposes. In the 1950s and early 1960s, US carriers were assigned strategic bombing duties, not least the use of a nuclear strike capability against the Soviet Union, but in the Korean and Vietnam Wars they were used extensively to provide ground support. The first major use of aircraft carriers was in the Korean War (1950–3). The carrier aircraft of the American Task Force 77 were of operational and tactical value, not least for ground support.

Carrier aircraft again played a major role in the Vietnam War, when they provided a nearby safe base for operations over both North and South Vietnam. Improvements in supply methods since World War II, for example re-supply from other ships, ensured that carriers were able to stay at sea for longer. During most of 1972, no fewer than six US carriers were on station off Vietnam, and, that summer, an average of 4,000 sorties were flown monthly.

In the Falklands War of 1982, Britain lacked a large aircraft carrier, and therefore airborne early warning of attack. They did, however, have two anti-submarine carriers equipped with Sea Harrier short take-off fighter-bombers armed with Sidewinder AIM-L missiles. These enabled Britain to contest Argentine air assaults on the task force and, in turn, to attack the Argentinians on the Falklands. Thanks to the carriers, which the Argentinians were not able to sink, the British had vital air support (but not superiority) for both sea and land operations.

The continued importance of carriers in the future is indicated by their role in recent American operations in the Middle East and by Britain's plans to build two super-carriers.

Submarines
World War II

'The only thing that frightened me
during the war was the U-Boat peril.'

WINSTON CHURCHILL

HAVING PLAYED A MAJOR ROLE IN WORLD WAR I, the potential use of submarines in the future excited considerable interest among naval planners. In addition there were important advances in the specifications of submarines, although their restricted sub-surface speed remained a major limitation.

This can be seen in the case of American submarines. The S class of 1918–21, with a range of 5,000–8,000 miles at a surface speed of 10 knots, was replaced by the B class: 12,000 miles at 11 knots. They were followed by the P boats of 1933–6, the first American submarines with a totally diesel-electric propulsion, and then by the Gatos introduced in 1940: double-hulled, all-welded-hull submarines with a range of 11,800 miles and a surface speed of 20–25 knots. By the time of the Japanese attack on Pearl Harbor in 1941, the American navy had 111 submarines in commission, while the Japanese had 63 ocean-going submarines.

The Japanese Type 1-400 submarines had a range of 37,500 nautical miles and a surface speed of 18.7 knots, and carried supplies for 60 days. In the event of war with the USA, the Japanese planned to use their submarines to hit American warships advancing into the western Pacific. They therefore intended to employ their long range as a major preliminary component in fleet action. In the event, their submarines failed to fulfil expectations, which was unsurprising as exercises in 1939 and 1940 had indicated significant deficiencies.

Ironically, the size of a submarine fleet was not a good indication of its importance during World War II. At the outset of the war, the Soviet Union had the largest submarine fleet and Italy the second largest. Neither, however, made much of an impact. In contrast, partly due to Hitler's interest in battleships, Germany only had 57 submarines.

CLEAR DETERMINATION

The most successful submarine campaign during the war was that by the Americans in the Pacific, and this became the most victorious one in history. American submarines proved effective and able to operate at long range. They had good surface speed and range. The Americans benefited from their ability to decipher Japanese signals, and from a clear determination to attack: unrestricted submarine warfare had been ordered after Pearl Harbor. The Japanese did not inflict enough casualties to cause a deterioration in American submarine leadership. The Americans were also helped by Japanese failure at convoy protection and anti-submarine warfare. The Japanese devoted insufficient resources to them, not least through not providing adequate air cover.

American submarines sank 5.32 million tons of Japanese merchant shipping (1,114 Japanese merchantmen), and forced the Japanese to abandon many of their convoy routes in 1944. The Japanese failed to build sufficient ships to match their losses, their trade was dramatically cut, and the imperial economy was shattered. There was no area of Japanese overseas and coastal trade that was free from attack. This greatly increased the uncertainty that sapped the predictability that industrial integration depended on. Japanese losses made it difficult to move raw materials in the conquered areas, such as iron, oil and rubber, back to Japan. This was crucial in thwarting plans to increase munitions production, as Japanese industry was dependent on the import of raw materials.

It was also difficult to move troops rapidly within the empire. The flexibility seen in the initial Japanese conquests had been lost. Expedients, such as postponing the maintenance of shipping, and leaving garrisons to fend for themselves, proved disastrous for overall effectiveness. By the summer of 1945, the naval war in the Pacific had been decisively won, and American submarines operated with few difficulties in the Yellow and East China Seas and the Sea of Japan.

GERMAN THREAT

The German campaign was less successful, but was still serious and sustained, and far more so than attacks by surface shipping or air attack. Submarines were less vulnerable than surface ships to blockade, detection and destruction, and could be manufactured more speedily and in large quantities. However, owing to an overlong commitment to surface warships, the Germans did not focus their entire naval construction effort on submarines until the spring of 1943. This only occurred after the German failure, on 31 December 1942, in the Battle of the Barents Sea, to use surface ships against less-heavily gunned British cruisers to block the British supply route to the Soviet Union, which led Hitler to replace Admiral Raeder as his naval commander by Admiral Dönitz. The latter was more committed to submarines.

Submarines were more sophisticated than in World War I, and, from 1940, the Germans had Norwegian and French bases, both of which they had lacked in the previous war. As a result, shipping sunk by U-boats rose from the summer of 1940, at a time when British warships were focused on home waters to cover the evacuation of forces from France and to retain control of the English Channel in the face of German invasion preparations. There were also severe losses in the winter of 1940–1, as the U-boats attacked Atlantic convoys and developed wolf-pack or group tactics.

Aircraft forced the U-boats to submerge, as a result of which their speed was slower and it was harder to maintain visual contact with targets, but the RAF, which was interested in strategic bombing and theatre fighters instead, had not devoted sufficient preparation to air cover for convoys against submarines. The demands of the bomber offensive against Germany on available long-range aircraft restricted the numbers available for convoy escort, while, due to limited range, land-based aircraft faced an 'Air Gap' across much of the mid-Atlantic.

A lack of sufficient preparation was also true of the British navy, which was primarily concerned with hostile surface shipping. The navy was content to rely on convoys and ASDIC (sonar) to limit submarine attempts, but the effectiveness of ASDIC was limited as

Submarines: World War II

1939
Within hours of war being declared, a German U-boat (from the German word *unterseeboot*) launches an attack on a liner and unrestricted submarine warfare begins

The U-29 sinks the air carrier HMS *Courageous*

HMS *Royal Oak* is sunk at anchor in a daring raid in Scapa Flow by U-47 with the loss of over 800 men. The German commander is awarded the Iron Cross by Hitler

1940
German U-boats are responsible for the loss of 2,606,000 tons of merchant shipping

1941 November
HMS *Ark Royal* is attacked by U-81 and sunk

1943
USS *Block Island* is sunk off the Canary Islands by U-549

1944
American submarines sink 5.32 million tons of Japanese merchant shipping

submarines preferred to hunt on the surface. The British had used carrier-based planes against submarines at the start of the war, but the sinking by U-29 of the carrier HMS *Courageous* on 17 September 1939 ended this practice, and the remaining fleet carriers were needed for operations against German and, from June 1940, Italian surface warships. Moreover, it took time to build escort aircraft carriers.

American entry into World War II against Germany in December 1941 was followed by very heavy losses of merchantmen to U-boat attacks in American coastal waters in the first half of 1942, but, in May, the situation improved considerably as effective convoying was introduced by the Americans. Moreover, the strategic failure of the U-boat offensive was further exposed by Allied, especially American, shipbuilding. In the first quarter of 1943, the Allies built more ships than the U-boats sank, and by the end of the third quarter they had built more than had been sunk since the start of the war. The war in the Atlantic was central to the eventual ability to apply American power against Germany, as it ensured that American troops could be transported across the Atlantic.

During the war, the Germans made major improvements to their submarines. In early 1944, they fitted *Schnorchel* devices which allowed the submarines to charge their batteries while submerged, as well as enabling the underwater starting and running of diesel engines, which reduced their vulnerability to Allied air power. The U-boats were becoming true submarines as opposed to just submersibles. But this did not significantly increase attack capability because the improvement in Allied escort effectiveness outweighed submarine advances. As a result, the Germans sank relatively few ships in 1944–5. Furthermore, the production of new types of submarine, especially the high-speed Type XXI Electro, was badly affected by Allied bombing. Had they been available in 1943, the situation would have been more serious for the Allies.

After World War II, there was little initial emphasis on submarines or anti-submarine capability, in large part because the USA dominated the oceans and at sea, stressed the potential of carriers for air attack on the Soviet Union. This show of force , however, was challenged by the build-up of the Soviet navy that began in the 1950s under Admiral Sergei Gorshkov, which quickly made the Soviet Union the world's number two naval power. Their Northern Fleet based at Murmansk, became their largest fleet, with a particularly important submarine component – about 400 strong by 1956.

This growth of the Northern Fleet obliged NATO powers to develop nearby patrol areas for submarines, as well as underwater listening devices, and also to create a similar capability in the waters through which Soviet submarines would have to travel *en route* to the Atlantic, both in the Denmark Strait between Iceland and Greenland, and between Iceland and Britain. Submarines also became of importance with the development of submarine-launched intercontinental ballistic missiles, which made them the principal naval vessel. The Americans led in this but others followed. Six Typhoons, the largest ballistic missile submarines built, entered Soviet service from 1980.

The USA feared that Soviet submarines would attack their trade routes, or launch missiles from near the American coast. This led the USA to focus on planning for naval conflict with the Soviets, rather than amphibious operations: the emphasis now was on being able to destroy Soviet naval power in battle and in its home waters. This resulted in an American emphasis on large attack submarines demonstrating that the type of submarine developed reflected the range of tasks that they could be expected to undertake.

Radar

'Radar is a method of using radio waves to detect the existence of an object... the word is coined from the initial letters of the phrase Radio Detection And Ranging.'

RAF STANDARD TECHNICAL TRAINING NOTES

RADAR WAS A PRIME EXAMPLE OF THE CRUCIAL CONTRIBUTION OF SCIENTIFIC ADVANCES TO THE CONDUCT OF WAR. Before and during World War II, radar, with its capacity for long-distance detection of movement, rapidly developed. Such detection was vital because of the greater range and speed of military units. As a result, reliance on human observers was inadequate, although the value of this system was increased by the use of telephone links with observers. Radar was used for naval and air capability and radar sets were installed in British warships from 1938.

It played an even more important role thanks to the integration of radar into the civilian defence network in Britain such as the rather crude Chain Home stations, which played a major role in helping the defenders against German air attack in the Battle of Britain in 1940. This advance was followed by the discovery of the sophisticated cavity-magnetron in 1940, which was central to the development of microwave radar.

In the Battle of Britain, radar not only helped the British, but also indicated a huge failure on the part of the Germans to understand British capability. This crucially was not only their lack of appreciation of the importance of radar but also its place in the integrated air-defence system. The Battle of Britain was a key episode both in British military history and in the history of air power.

The Battle of Britain was the first major check experienced by the Germans, and one that was critical to the survival of Britain as an independent state, for the air attack was designed to prepare the way for Operation Sealion, the planned German invasion, particularly by driving British warships from the English Channel. British victory reflected both the deficiencies of the numerically superior German air force, and the capabilities and fighting quality of its numerically inferior British opponents.

There were also serious problems with German strategy. The Luftwaffe (German air force) had been ordered to drive British warships from the English Channel, but its commanders were increasingly focused on attacking the RAF (Royal Air Force) and its supporting infrastructure, in order to prepare the way for reducing Britain to submission by a bombing war on civilian targets, a strategy that would put the Luftwaffe centre-stage, as its head, Reichsmarschall Hermann Göring, intended.

The lack of clarity in the relationship between air attack and invasion affected German planning, but there was also a lack of preparation for a strategic air offensive, because the Germans had not sufficiently anticipated its necessity. Furthermore, the schedule was too tight, with the need to invade before mid-September, when the weather was likely to become hostile. In addition, the Germans suffered in 1940 from a lack of properly trained pilots, and from limitations with their planes and tactics. Their bombers' load capacity and range were too small, and their fighters were too sluggish (Me-110) or had an inadequate range affecting the time they could spend over England (Me-109), and the fighters were also handicapped by having to escort the bombers. Furthermore, they had lost many planes and pilots during the Battle of France.

Despite these deficiencies, the Germans outnumbered the British on the eve of battle. There had been heavy losses of British planes in the Battle of France earlier that summer, and the British factories where new planes were being manufactured were now within range of German bombers.

Initial German attacks on the RAF and its airfields inflicted serious blows, particularly on pilot numbers, and it was a shortage of pilots, particularly trained pilots, rather than of planes, that threatened Britain's survival. There was an acute crisis in late August and early September 1940. However, fighting over Britain, it was possible to recover pilots who survived being shot down, while the RAF also benefited from the support provided by radar and the ground-control organization, as well as from able command decisions, good intelligence and a high level of fighting quality. Although German radar in 1940 was technically superior to the British variety, the integrated air defence system greatly helped the British defence. Given that it had been set up within four years, it was an impressive achievement.

Radar, however, faced difficulties, not least that of providing sufficient information rapidly. In particular, it was hard to predict the number and height of attacking planes, while it took four minutes for radar information to reach the British fighter squadrons. The Germans also used feints and diversions to confuse the British.

THE BLITZ

The Germans switched, in early September, to bomb London and other cities in the Blitz. In part, this shift occurred because of faulty intelligence and, to a minor extent, in response to a British raid on Berlin on the night of 25–6 August. The Luftwaffe hoped that these attacks would force the British to commit their reserves, but in fact it led

Radar

1887
Heinrich Hertz, the German physicist discovers that radio waves penetrate different materials
1904
Christian Hülsmeyer first uses radio waves to detect the presence of distant metallic objects
1917
Nikola Tesla in *The Electrical Experimenter* sets down criteria for basic radar
1934
The French CSF company takes out a patent for detecting obstacles by ultra-short wavelengths
1935
Robert Watson-Watt publishes *The Detection of Aircraft by Radio Methods.* The success of the Daventry Experiment leads the Air Ministry to develop radar in the UK
1936
The US Naval Research Laboratory demonstrates pulse radar successfully
1941
Ground Control Intercept Stations are set up in the UK
1941
The term RADAR is coined as an acronym for Radio Detection and Ranging

to the Luftwaffe being heavily pummelled from 7 to 15 September. Furthermore, as a result of the change of policy, the pressure on the RAF infrastructure had diminished, although the strain on the population was heavy. The Germans were out to destroy civilian morale, but, although there were occasional episodes of panic, on the whole it remained high.

In turn, radar played a role in the air defence of Germany against British and American air attacks, which gathered pace in 1942 and became large scale in 1943. The Germans developed a complex and wide-ranging system of radar warning, with long-range, early-warning radars, as well as short-range radars that guided night fighters, which also had their own radars, towards the bombers. These enabled the Germans to inflict very heavy losses on Allied bombers.

Radar-defence systems, however, could be wrecked by the British use of 'window': strips of aluminium foil that appeared like bombers on the radar screens. In response, from the summer of 1943, the Germans relied on radar guidance to the general area of British air activity, which 'window' contributed to, and visual sightings thereafter. This caused British aircraft losses to mount from the late summer of 1943. Furthermore, that autumn, German radar was adapted so as to be able to circumvent the impact of 'window'. Electronic warfare, in which attackers and defenders strove to outwit each other, was increasingly evident.

NAVAL OPERATIONS

Radar was also used in naval operations. In May 1941, when the German battleship *Bismarck* was trying to break into the North Atlantic in order to hit British supply links, particularly the crucial link with the USA, it was unclear which route she would take. British bombers had failed to find the *Bismarck* in Norwegian waters, but she was spotted by patrolling British warships in the Denmark Strait, between Iceland and Greenland, on 23 May. The following day, when the *Bismarck* encountered a British squadron sent to intercept her, ship radar helped the British shadow the German warships, although, in the subsequent gunnery exchange, the *Bismarck* sank the battle cruiser *Hood* (only three of the crew of 1,418 survived) and seriously damaged the battleship *Prince of Wales*. A shell from the latter, however, hit the *Bismarck*, causing a dangerous oil leak that led the commander to set course for France and repairs. She was sunk before she could reach her destination.

In the Pacific, despite extensive research, the Japanese lacked the Americans' advantage in radar, an advantage that proved very important in naval conflict there. On 14 November 1942, off the island of Guadalcanal, the radar-controlled fire of the battleship *Washington* pulverized the Japanese battleship *Kirishima*. In covering the landing on Bougainville in the Solomon Islands on 1 November 1943, a force of American cruisers and destroyers beat off an attack that night by a similar Japanese squadron, with losses to the latter in the first battle fought entirely by radar.

Radar also helped American carriers defend themselves against air attacks, as in the Battle of the Philippine Sea in 1944 in which Japanese air attacks were located by American radio. In early 1945, in defending against *kamikaze* attacks, the Americans benefited from the large number of fighters carried by their numerous carriers and from the radar-based system of fighter control. Escort vessels and aircraft in the fight against submarines also benefited from radar. Microwave radar was the most effective counter to the U-boat, because the Germans were not able to develop suitable countermeasures in time. The Germans experienced a major shock when they discovered that the British were using microwaves. Until then, German scientific orthodoxy was sceptical of their value, and no priority had been given to the production of the transmitting valves necessary for microwaves.

Radar also revealed a vital difference in scientific cultures between the two sides. In contrast to the teams of scientists in Britain and the US, who were free to handle their own affairs, German technical staff were frightened of being proved wrong, or exposed to ridicule. Authoritarian political and administrative direction in Germany proved counterproductive.

After the war, the development of strategic bombing forces capable of launching nuclear attacks led to a renewed emphasis on radar defences. The threat from the Soviet Union across the North Pole led the US to develop radar systems in Canada: the Pinetree network in 1954, and the Distant Warning Line in 1957. There was also the all-Canadian Mid-Canada Line in 1957. These systems, however, were not designed to cope with intercontinental ballistic missiles which were to pose new problems for detection, particularly those of speed.

Aircraft
World War II

'If we lose the war in the air, we lose the
war, and we lose it quickly.'

FIELD MARSHAL VISCOUNT MONTGOMERY OF ALAMEIN

THE 1930s SAW A MARKED IMPROVEMENT IN THE FLYING STANDARDS AND COMBAT CHARACTERISTICS OF AIRCRAFT. This was particularly so for fighters in the mid- and late 1930s, as wooden-based bi-planes were replaced by all-metal cantilever-wing monoplanes with high performance engines capable of far greater speeds, for example the American P-36 Hawk, German Me-109 and Soviet I-16 Rata. The British developed two effective and nimble monoplane fighters, the Hawker Hurricane and Supermarine Spitfire. Alongside early-warning radar, they were to help Britain resist the German air onslaught in 1940.

In the 1930s, the range and armament of fighters, and the range, payload and armament of bombers all increased. In Britain, there was strong concern about the likely impact of the German bombing of civilian targets. The major impact on public morale of German raids on London in World War I seemed a menacing augury. It was believed, in the words of the ex- and future Prime Minister Stanley Baldwin in 1932, that 'the bomber will always get through'.

TERROR BOMBING

Air attack was central to the German offensives of the early stages of the war. Ground-support dive-bombing was valuable, especially in Poland (1939), France (1940) and Greece (1941), while the terror bombing of cities, for example Warsaw in 1939, Rotterdam in 1940 and Belgrade in 1941, was seen as a way to break the will of opponents. However, the inadequately prepared Germans were outfought in the sky when they attacked Britain in 1940. Similarly, Axis air power had serious deficiencies in affecting the war at sea. Against the Soviet Union on the East Front, although German air power was of tactical value, it lacked the capability to achieve important strategic goals.

During World War II, alongside failures in execution, there were also major advances in air capability. In part, this was a matter of aircraft, although there was also an improvement in such spheres as ground support. The training of large numbers of aircrew was a formidable undertaking, but it paid off, particularly for the Allies.

For example, in the Pacific, there was a growing disparity in quality between American and Japanese pilots, a matter of numbers, training and flying experience. As a result, the Japanese could not compensate for their growing numerical inferiority in the air. At the same time, it would be foolish to overlook the extent to which the Americans by 1943 benefited in the Pacific from better aircraft. Whereas the Japanese had not introduced new classes of planes, the Americans had done so, enabling them to challenge the Zero fighter which had made such an impact in the initial Japanese advances. The Corsair, Lightning and Hellcat outperformed the Zero, while, as their specifications included better protection, they were able to take more punishment than Japanese planes. The Japanese had designed the Zero with insufficient range and manoeuvrability, because the safety of their pilots was a low priority.

The provision of improved aircraft was also important to the Anglo-American air offensive against Germany. On 19 May 1943, Winston Churchill, the British Prime Minister, noted, in an address to a joint session of the US Congress, that opinion was 'divided as to whether the use of air power could, by itself, bring about a collapse of Germany or Italy. The experiment is well worth trying, so long as other measures are not excluded.' Despite the limited precision of bombing by high-flying planes dropping free-fall bombs, strategic bombing turned out to be crucial to the disruption of German logistics and communications, largely because the bombing was eventually on such a massive scale. An article in *The Times* of

1 May 1945, significantly entitled 'Air Power Road to Victory... 1939 Policy Vindicated', claimed that reductions in oil output due to air attack had affected German war potential in all spheres, 'neither his air force nor his army was mobile'. Indeed, the German oil system had been deliberately targeted.

More generally, area (rather than precision) bombing disrupted the German war economy, although it also caused heavy civilian casualties, notably at Hamburg in 1943 and Dresden in 1945, and by 1943, Anglo-American bombing had wrecked 60 per cent of Italy's industrial capacity and badly undermined Italian morale. Air attack on Germany also led to the Germans diverting much of their air force and anti-aircraft capacity to home defence, rather than supporting frontline units, and also to an emphasis on the production of anti-aircraft guns.

The civilian casualties inflicted by Allied bombing have since become a matter of controversy, but too little attention has been devoted to the expectations, from both domestic opinion and the Soviet Union, that major blows would be struck against Germany prior to the opening of the 'Second Front' by means of an invasion of France.

Fighter aircraft of World War II

1936 The prototype of the Supermarine Spitfire makes its first flight. 310 aircraft are ordered

1937 Early versions of the Messerschmitt Bf-109 see service with the German Condor Legion during the Spanish Civil War

1940 The Hurricane forms the bulk of fighters taking part in the Battle of Britain accounting for 60 per cent of claimed German losses. But it is the Spitfire which is best able to compete with the newest German Bf-109s

The Mitsubishi Zero enters service with the Japanese navy quickly establishing itself as the best carrier-based fighter in use at the time

1941 The Focke-Wulf Fw-190 enters service with the Luftwaffe establishing superiority over current versions of the Spitfire. 20,000 are produced

The first twin-engined Lockheed P-38 Lightning fighters are delivered to the US Air Corps

1942 The RAF test the new American Mustang fighter

1943 The arrival of the long-range, Rolls Royce Merlin-powered Mustang P51B in Europe transforms the air war ensuring air superiority over Germany

1944 The Messerschmitt Me-262 jet fighter appears over Europe but cannot influence the outcome of the war

The delay of this invasion from first 1942 and then 1943, led to pressure for action. This matched the pressure on the Western Allies in World War I to mount attacks in 1915 and 1916, in order to reduce the load on Russia, pressure that led both to the Gallipoli operation of 1915 and to offensives on the Western Front. Also the Germans had not only begun the bombing of civilian targets, but, with the coming of the V-Is in 1944, also launched missiles against British cities.

However, especially prior to the introduction of long-range fighters, bombers were very vulnerable. Cripplingly heavy casualty rates occurred in some raids, for example those of the American 8th Air Force against the German ball-bearing factory at Schweinfurt in August and October 1943. The majority of the bombers were lost to German fighters, with anti-aircraft fire and accidents occurring for the rest. British night attacks on Berlin from 18 November 1943 until 31 March 1944, which it had been promised would undermine German morale, led instead to the loss of 492 bombers, a rate of losses that could not be sustained. In the British raid on Nuremberg of 30–31 March 1944, 106 out of the 782 bombers were lost, with only limited damage to the city and few German fighters shot down. This led to the end of the bomber-stream technique of approaching the target.

Strategic bombing was made more feasible by four-engined bombers, such as the British Lancaster and the American B-29, as well as by heavier bombs and developments in navigational aids and training. Heavily armed bomber formations lacking fighter escorts proved less effective in defending themselves than had been anticipated. This made the introduction of long-range fighter escorts important, especially the American P-38s (Lightnings), P-47s (Thunderbolts), and P-51s (Mustangs). Both the latter used drop fuel tanks which could be jettisoned when used up. The Mustangs, of which 14,000 were built, were able not only to provide necessary escorts but also, in 1944, to seek out German fighters and thus win the air war above Germany. This contrasted with the Luftwaffe's failed offensive on Britain in 1940–1.

The Mustang's superiority to German interceptors was demonstrated in late February and March 1944, when major American raids in clear weather on German sites producing aircraft and oil led to large-scale battles with German interceptors. Many American bombers were shot down, but the Luftwaffe also lost large numbers of planes and pilots. The latter were very difficult to replace, in large part because training programmes had not been increased in 1940–2, as was necessary given the scale of the war, and this helped to ensure that, irrespective of aircraft construction figures, the Germans would be far weaker. Towards the end, the Germans could not even afford the fuel for training, while a lack of training time was also a consequence of the shortage of pilots. In 1943, the Allies did not yet have sufficient air dominance to seek to isolate an invasion zone, but, by the time of the Normandy landings in June 1944, the Germans had lost the air war.

AIR ASSAULT ON JAPAN

By late 1944, the air assault on Japan was gathering pace. Initially, the American raids were long-distance and unsupported by fighter cover. This led to attacks from a high altitude, which reduced their effectiveness. The raids that were launched were hindered by poor weather, especially strong tailwinds, and difficulties with the B-29's reliability, as well as the general problems of precision bombing within the limited technology of the time.

From February 1945, there was a switch to low-altitude night-time area bombing of Japanese cities. The impact was devastating, not least because many Japanese dwellings were made of timber and paper and burned readily when bombarded with incendiaries, and also because population density in the cities was high. Fighters based on the island of Iwo Jima (three air hours from Tokyo) from 7 April 1945 could provide cover for the B-29s, which had been bombing from bases on the more distant island of Saipan since November 1944.

Weaknesses in anti-aircraft defences, both planes and guns, eased the American task and made it possible to increase the payload of the B-29s by removing their guns. Although the Japanese had developed some impressive interceptor fighters, especially the Mitsubishi AbM5 and the Shiden, they were unable to produce many due to the impact of Allied air raids and of submarine attacks on supply routes, and were also very short of pilots. In 1944–5 American bombers destroyed over 30 per cent of the buildings in Japan, including over half of Tokyo and Kobe. The deadliness of bombing was amply demonstrated.

Landing Craft

'And when the sun rose the next morning, I saw the invasion fleet lying off the shore. Ship beside ship. And without a break, troops, weapons, tanks, munitions and vehicles were being unloaded in a steady stream.'

A German private on D-Day, 6 June 1944

POWERED LANDING CRAFT MADE A GREAT DIFFERENCE TO THE SAFETY OF CREW in amphibious operations. They were built in World War I, with the British doing so from 1914, but most of the troops that went ashore in such operations, for example at Gallipoli in 1915, were landed from ordinary ships. This meant steam-driven vessels that could not beach, and, as a result, troops were often landed into dinghies, a vulnerable situation, or into shallows that were far from shallow. Equipment had to be landed at a port.

There was more interest in the interwar period in specialized landing craft, particularly by the USA, but the Japanese made the most progress, both in developing types and in building a reasonable number of ships. Their Dai-Hatsu had a ramp in its bows and this was to become the key type of landing craft.

AMPHIBIOUS OPERATIONS

World War II saw the much greater development of specialized landing craft and a major expansion in amphibious operations. The military writer J.F.C. Fuller pointed out that Operation Overlord, the landing in Normandy on D-Day, 6 June 1944, registered a key transformation in amphibious operations as there was no need to capture a port in order to land, reinforce and support the invasion force. He wrote in the *Sunday Pictorial* of 1 October 1944,

> had our sea power remained what it had been, solely a weapon to command the sea, the garrison Germany established in France almost certainly would have proved sufficient. It was a change in the conception of naval power which sealed the doom of that great fortress. Hitherto in all overseas invasions the invading forces had been fitted to ships. Now ships were fitted to the invading forces... how to land the invading forces in battle order... this difficulty has been overcome by building various types of special landing boats and pre-fabricated landing stages

To Fuller, this matched the tank in putting the defence at a disadvantage. The preliminary attack on the French port of Dieppe on 19 August 1942, in effect a small-scale practice for the Allied landing on D-Day, had shown that attacking a port destroyed it. This assault on a well-fortified position led to heavy Allied, mainly Canadian, casualties, to accurate machine-gun, artillery and mortar fire, and underlined the problems of amphibious attacks on a port. In 1944, instead, the Allies decided to bring two prefabricated harbours composed of floating piers with the invasion, although to achieve the invasion it was necessary to gain control of ports such as Antwerp and Marseilles to deal with logistical problems.

By then, the British and the Americans had totally taken the lead in the development and production of landing craft. Aside from the general landing craft with bows that became ramps, new specialized landing craft were produced that could land vehicles. In 1940, the British launched their first Landing Craft Tank (LCT), and this proved a very versatile vessel as it was able to land much more than just troops alone.

There was also a marked improvement in landing craft for troops. For example, the American Marines first used Higgins boats, also known as Eurekas, which were wooden infantry landing craft. Constructed at first for civilian use, they could be easily withdrawn from a beach, but their wooden hulls proved vulnerable and they were replaced in 1943.

In 1944, the Germans still, mistakenly, anticipated that the Allies would focus on seizing ports. Instead, the emphasis was on the beaches, with 2,470 landing craft used in the assault phase supported by other specialized fighting vehicles. These included Duplex Drive (amphibious) Sherman tanks, as well as the tanks developed by the British to attack coastal defences, for example Crab flail tanks for use against minefields.

OPERATION TORCH

Earlier experience with amphibious landings also proved important. Operation Torch – the largely American invasion of North Africa in 1942 – had been followed by Anglo-American invasions of Sicily and mainland Italy in 1943. The success of the Torch landings was only a limited indicator of capability, as opposition to them had been weak. The landings of 1943, in contrast, like that at Anzio in Italy in January 1944, provided warnings about the difficulty of invading France, especially in terms of the strength of the German response.

The Americans also acquired considerable experience of amphibious operations against the Japanese in the Pacific. Powered landing craft ensured that it was possible to attack defended coastlines simultaneously, which greatly limited options for the defenders. The British and Americans gained important cumulative experience in successful amphibious operations, including the use of landing craft, and in their co-ordination with naval and air support, not least in air support and airborne attacks. They developed and used a variety of effective specialized craft, including tracked landing vehicles. However, the difficulties that were to be faced by the Americans on Omaha Beach on D-Day showed that the lessons learned in the Pacific did not apply everywhere. Some that were, however, were not applied in Europe, in large part due to inter-service and inter-theatre rivalries, while there were also the particular problems posed by German defensive ideas and fortifications.

On D-Day, 6 June 1944, the Allies were also helped by the German response. The Germans lacked adequate naval and air forces to contest an invasion, and much of their army in France was of indifferent quality and short of transport and training, and, in many cases, equipment. The commanders were divided about where the attack was likely to fall and how best to respond to it, particularly over whether to move their ten panzer (tank) divisions close to the coast, so that the Allies could be attacked before they could consolidate their position, or if they should be gathered in a mass as a strategic reserve. The eventual decision was for the panzer divisions, whose impact greatly worried Allied planners, to remain inland, but their ability to act as a strategic reserve was lessened by choosing not to mass them and by Allied air power. This decision reflected the tensions and uncertainties of the German command system.

> Iwo Jima and Okinawa were seized in order to provide airbases for an attack on Japan. The flexibility provided by landing craft ensured that the key fighting occurred once the troops were landed. Once so, there was bitter combat on land. Heavy casualties were suffered in defeating the well-positioned Japanese forces, who fought to the death with fanatical intensity for islands seen as part of Japan.

SURPRISE LANDINGS

A successful Allied deception exercise, Operation Fortitude, ensured that the Normandy landing was a surprise. The Germans had concentrated more of their defences and forces in the Calais region, which offered a shorter sea crossing and a shorter route to Germany. Normandy, in contrast, was easier to reach from the invasion ports on the south coast of England, particularly Plymouth, Portland and Portsmouth.

In 1945, the focus on amphibious operations returned to the Pacific. The key amphibious operations in 1945, all mounted by the Americans, were on Luzon, the principal island in the Philippines, and on the islands of Iwo Jima and Okinawa. These were only the most prominent of a number of operations, particularly in the extensive Philippine archipelago. They were required in order to ensure the security of shipping passages, and there was a series of amphibious attacks. Samar and Palawan were invaded in February, and other islands in March and April. American skill in amphibious operations, in the co-ordination of air and sea support, and in the rapid securing and development of bridgehead positions, was demonstrated and enhanced. This continual process was a useful preparation for the planned landings in the Japanese archipelago.

Iwo Jima and Okinawa were seized in order to provide airbases for an attack on Japan. The flexibility provided by landing craft ensured that the key fighting occurred once the troops were landed. Once so, there was bitter combat on land. Heavy casualties were suffered in defeating the well-positioned Japanese forces, who fought to the death with fanatical intensity for islands seen as part of Japan. The Japanese were also skilful defenders, well able to exploit the terrain, not least by tunnelling. This ensured that the bombing and shelling that preceded the American landing inflicted only minimal damage. As a consequence, the conquest of the islands was slow and bloody, and much of the fighting was at close quarters.

At the close of the war, plans were far advanced for a British landing on the coast of Malaya in 1945 and an American landing on that of Japan in 1946, but both were shelved thanks to the use of the atomic bomb.

THE KOREAN WAR

The most important contested landing after World War II, occurred at Inchon on the west coast of Korea in September 1950, a key moment in the Korean War. In a daring and unrehearsed landing, far behind the front, about 83,000 troops were successfully landed in difficult, heavily tidal waters. They pressed on to capture nearby Seoul, wrecking the coherence of North Korean forces and their supply system. This achieved a major psychological victory, and was followed by the driving back of North Korean forces to near the Chinese border.

After this there was less use of landing craft. In large part, this was because invasions, such as that of the Suez Canal zone in Egypt, in the Suez Crisis of 1956, increasingly used helicopter-borne troops and parachutists. This was the case in the Allied attack on Iraq in 2003 where the emphasis was on attacking the littoral rather than simply the coast. This meant that the Allies projected strength deep into the interior and air power had greater significance than amphibious capability.

Jet Aircraft

'The idea that superior air power can in some way be a substitute for hard slogging and professional skill on the ground... is beguiling but illusory.'

AIR MARSHAL SIR JOHN SLESSOR,
AIR POWER AND WORLD STRATEGY OCTOBER 1954

THE JET AGE HAD A SLOW BEGINNING and due to lack of British government funding, Allied jet fighters arrived in service too late to affect the course of World War II. As early as 1930, Frank Whittle, a British air force officer, patented the principles that led to the first gas turbine jet engine which he first ran under control conditions in 1937. His brilliant innovation was rapidly copied and the Germans in 1939 and the Italians in 1940 beat the British jet into the air. In 1944, the British brought Meteors (capable of 490 mph/788 km) and the Germans the Messerschmitt Me-262 into service. The Allies found that the speed of the Me-262 (540 mph/870 km) made it difficult to tackle and its tactics posed serious problems for the Allies. It would dive at high speed through the Allied fighter screen and continue under the bombers prior to climbing up in order to attack the bombers from behind.

If, however, the Me-262 was involved in a dogfight, it was vulnerable as it had a poor rate of turn. There were also efforts to catch it when even more susceptible: on take-off and landing.

The Germans had insufficient numbers of the Me-262 to transform the course of the war, and its late entry into the war was also significant. The Germans had only focused production on the Me-262 after considerable delay, in part because Professor Messerschmitt was also keen to continue work on his projected Me-209, a conventional piston-engined plane. There was also delay because of interest in the use of the plane as a bomber. As well, there was separate work on other jet planes, the Arado Ar-234, which was designed as a jet bomber and reconnaissance aircraft, and the Ju-287, a four-engined jet bomber.

Allied air raids caused delay and problems such as a shortage of fuel, exacerbating the serious problems in the German economy caused by poor organization and the mismatch of goals, systems and resources. There was a shortage of raw materials that led to difficulties with blade fractures in the turbine rotars. This was a more serious drawback than Hitler's views on the use of the plane which were partly responsible for squandering the German lead in jet-powered aircraft because of his preference that the Me-262 should not be used as an interceptor of Allied bombers, despite its effectiveness in the role but rather as a high-speed bomber. Indeed, in June 1944, he ordered its name changed to *Blitzbomber*. Only 564 Me-262 were built in 1944. Furthermore, the plane had problems, both with the engines and an inadequate rate of turn.

POST-WAR DEVELOPMENT

Jet aircraft developed rapidly after the war. The first successful carrier landing of a jet aircraft took place on HMS *Ocean* in December 1945. The Korean War (1950–3) saw the first dogfights between jet aircraft. The Chinese, who had only created an air force in November 1949, and whose Soviet-trained pilots lacked adequate experience and were equipped with out-of-date Soviet planes, were supported by the advanced MIG-15 fighters of the Manchurian-based Soviet 'Group 64'. These fought American F-86 Sabres in the air space over Korea. The rotation system employed greatly undermined the Soviet pilots' continuity of experience, and thus their effectiveness. The Americans inflicted far heavier casualties and were able to dominate the skies, with serious consequences for respective ground support, although the absence of an adequate and integrated command limited American exploitation of this advantage.

In the 1950s, jet fighter-bombers, such as the American F-84 Thunderjet, made their first appearance, and they came to play a major role, replacing more vulnerable World War II-period planes. The Americans also deployed long-range jet bombers (B-47s and B-52s), as well as jet tankers (KC-135s).

VIETNAM

The greater capability of jet aircraft led to the enhanced use of air power in the Vietnam War. Over half the $200 billion spent on the war, a sum far greater than that expended by other Western powers on decolonization struggles, went on air operations, and nearly eight million tons of bombs were dropped on Vietnam, Laos and Cambodia; South Vietnam became the most heavily-bombed country in the history of warfare. There were also major American bombing offensives against North Vietnam, which were designed to fulfil both operational and strategic goals: to limit Northern support for the war in the South, and to affect policy in the North by driving the North Vietnamese to negotiate.

These goals were not fully fulfilled but there was an enhanced effectiveness in American air power

Early jet aircraft

Frank Whittle (1907–96) and Hans von Ohain (1911–98) are the co-inventors of the jet engine, working with no knowledge of each other's progress

1930 Frank Whittle registers his patent for the turbo-jet engine but his ideas are not received with enthusiasm at the British Air Ministry. Whittle does not renew his patent which lapses in 1935

1936 Hans von Ohain is granted a patent for his turbo-jet engine

1939 The German jet-powered Heinkel HE178 has its maiden flight

1941 The English Gloster Whittle powered by the WI engine makes its maiden flight

1942 The prototype of the jet-powered Messerschmitt Me-262 makes its maiden flight

1943 The British Gloster Meteor makes its maiden flight

1944 The Messerschmitt Me-262 is the first true jet fighter to see active service.

The Gloster Meteor is reserved for operations against the unmanned German V-1 rockets

by 1972. This was due to a marked improvement in American air capability that reflected new thinking in military doctrine, stemming from adjustment to the varied needs of the Vietnam war, and the use of laser-guided bombs. These compensated for earlier limitations of bomb accuracy caused by flying at high altitudes above anti-aircraft fire.

Air power was also very important in the Arab–Israeli wars. This was demonstrated in 1967 when Israel launched a pre-emptive attack on Egypt in order to deal with the growing aggression of its ruler, Colonel Nasser. The Israeli assault began on 5 June with a surprise attack on the Egyptian air bases, launched by planes coming in over the Mediterranean from the west. The Egyptians, who had failed to take the most basic precautions, lost 286 planes in just one morning, and, in addition, their runways were heavily bombed. Gaining air superiority in this fashion proved crucial to the subsequent land conflict. Egyptian ground forces were badly affected by Israeli ground-support attacks. Jordan joined in that day on the Egyptian side, only to have its air force destroyed by the Israelis, and the West Bank was subsequently overrun.

MASTERY OF THE AIR

Air power proved important again when conflict resumed in 1973. In 1978, Israel advanced into southern Lebanon in an attack on the Palestine Liberation Organisation – the Israeli advance benefited from close air support. In 1982, Lebanon was invaded once more with the Israelis gaining the advantage over the rival Syrians who were established there. Again, air power proved important. The Syrians initially fought well, but, once their missile batteries in Lebanon had been knocked out, and their air force badly pummelled by Israeli aircraft armed with Sidewinder missiles and supported by electronic counter-measures, they proved vulnerable to Israeli attack, now bolstered by clear mastery in the air. Israeli air power proved less effective in Lebanon in 2006.

Jet aircraft also enhanced the auxiliary functions of air power, supply and reinforcement. Air power could be used to move large numbers of troops overseas more rapidly than ships. In response to disorder in the Dominican Republic in the spring of 1965, the USA airlifted 23,000 troops in less than two weeks. Considerable Soviet airlift capacity, in turn, was demonstrated in Angola in 1975. As a result of such action, airports became key points of operational importance, and seizing control of them became a crucial goal in coup attempts. Soviet troops were flown into Prague airport in 1968 when the Czech government was overthrown.

The effectiveness of air power, however, became a matter of considerable comment. The defeat of Iraq in 2001 was seen as a triumph for air power, but in 1999, the effectiveness of the major NATO air assault on Serbia that was designed to achieve a Serbian withdrawal from Kosovo was called into question. This assault suffered the loss of only two aircraft, but the subsequent Serbian withdrawal from Kosovo revealed that NATO estimates of the damage inflicted by air attack, for example to Serb tanks, had been considerably exaggerated. Benefiting from the limitations of Allied intelligence information, and its serious consequences for Allied targeting, and from the severe impact of the weather on air operations (a large number cancelled), the Serbs, employing simple and inexpensive camouflage techniques that took advantage of terrain and wooded cover, preserved most of their equipment, despite 10,000 NATO strike sorties. The air assault also revealed the contrast between output (bomb and missile damage) and outcome: the air offensive did not prevent the large-scale expulsion of Kosovars from their homes, and this actually increased as the air attack mounted. The Serb withdrawal may have been due more to a conviction, based in part on Russian information, that a NATO land attack was imminent. The crisis suggested that air power would be most effective as part of a joint strategy.

Furthermore, although the damage to the Serbian army from air attack was limited, the devastation of Serbia's infrastructure, in the shape of bridges, factories and electrical power plants, was important, not least because it affected the financial interests of the élite as well as their morale, and the functioning of the economy. This shows the marked contrast between the limited tactical, and possibly more effective strategic impact of air power.

American air power also played a major role in Afghanistan in the overthrow of the Taliban regime. The air attack helped switch the local political balance within Afghanistan. Impact analysis revealed that bombing was subsequently less effective in support of the ground operations near Tora Bora in December 2001 and in Operation Anaconda, east of Gardez the following March. This was attributed to the Taliban ability to respond by taking advantage of terrain features, for camouflage and cover. The differences between effort, output and outcome were amply demonstrated.

Tanks
World War II and After

'The tanks now rolled in a long column
through the line of fortifications and on
towards the first houses, which had been set
alight by our fire... engines roared, tank tracks
clanked and clattered... On we went at a
steady speed towards our objective.'

ERWIN ROMMEL ON CROSSING THE MAGINOT LINE, 15 MAY 1940

THE TANKS OF WORLD WAR II WERE SCARCELY THE SAME WEAPON AS THEIR PREDECESSORS of 1916–18. This was just as well as they were expected to engage in a very different war, one characterized by greater mobility, both tactically and operationally. Some of this had been predicted and planned for by interwar writers and planners, although it would be misleading to see the tank as the centre of military planning. Instead, much continued to revolve around infantry and artillery.

The Germans made much of their commitment to military mechanization, and created the first three panzer (armoured) divisions in 1935. These were designed to carry out the doctrine of armoured warfare that was developed in Germany, in particular by Heinz Guderian. Initially drawing heavily on Britain's use of tanks in World War I, and on subsequent British thought, especially that of J.F.C. Fuller, the Germans developed their own distinctive ideas in the late 1930s. They planned to use tanks *en masse* in order to achieve a deep breakthrough, rather than employing them, as the French did in support of infantry. The panzer divisions were to seize the initiative, to move swiftly, and to be made more deadly by being combined arms units incorporating artillery and infantry.

GERMAN *BLITZKRIEG*

Indeed, tanks were to be the cutting-edge of the successful 'lightning-strike' German *blitzkrieg* operations from 1939 to 1941, at the expense of Poland (1939), the Netherlands (1940), Belgium (1940), France (1940), Yugoslavia (1941) and Greece (1941). German panzer divisions proved operationally effective as formations which maximized the idea of the tank as a weapon. Handled well, as by the British in their counterattack near Arras on 21 May 1940, Allied tanks could also be effective, but, on the whole, the Germans controlled the pace of the armoured conflict, not least because their tank doctrine was more effective. The Germans were able to make the offensive work in both operational and tactical terms, amply displaying the potential, for both, of an attack spearheaded by armoured forces. Tanks could move across country, limiting the need to tie forward units to roads.

The French had more tanks than the Germans in 1940, and their tanks were, for the most part, more heavily gunned and had better protection. The best French tank, the Char B, had far thicker armour than its German counterparts, which were vulnerable to both tank fire and anti-tank guns. Many German tanks were very lightly armoured and gunned. However, French tanks were also somewhat slower and many had to turn in order to fire. Tanks with one-man turrets left the commander to be the gunner and the communicator. The French persisted in seeing tanks as a form of mobile artillery, and split them up accordingly into small groups, rather than employing the tanks as armoured divisions for their shock value.

In World War II, there were difficult trade-offs in tank warfare between speed, armour, armament, durability and ease of production. Unlike the Germans, the Americans and Soviets concentrated on weapons that made best use of their industrial capacity because they were simple to build, operate and repair, such as the American Sherman M-4 tank. In contrast, German tanks were complex pieces of equipment and often broke down. Much German armour was no better than that of the Soviets, and although the British and Americans had more tanks available for a long time there were problems with their effectiveness. The British Infantry Mark I, Valentine and Churchill tanks suffered from inadequate armament, and all bar the last were undergunned.

FIREPOWER AND ARMOUR

The best German tanks were technically better in firepower and armour in 1944, the Tiger and Panther, for example, being superior in both to the Sherman, but the Sherman was better than the Mark IV, which remained the most numerous type of German tank, while the unreliability and high maintenance requirements of the Tiger tank weakened it.

Tank tactics were also important. For example, by early 1942, the Soviets had impressive tanks in the T-34 and KV-1, but the Germans proved more effective in mobile warfare. The Red Army was developing armoured units at the level of tank armies in response to the German tanks, but, as yet, their own tank tactics did not match those of the Germans. In particular, the Soviets placed more of an emphasis on supporting infantry.

In 1943, in the Battle of Kursk, the largest tank battle in history, the German tank force was weakened as it fought its way through prepared Soviet defences, and the Red Army was then better able to commit its tank reserves. The Germans had Panther and Tiger tanks, but the standard Soviet T-34 had been upgraded, and was used effectively at close range, where it matched up well with the new German tanks. As Soviet production of tanks was far greater than that of Germany, they could better afford to take losses, and Kursk shifted the situation in their favour.

In 1944, the Soviets benefited from the continued improvement in their tanks to match new German tank types. The T-34/85 was more heavily gunned than its predecessors. From June 1944, the Soviets had much more success in pushing encirclements through to destruction.

The last stages of the war demonstrated the continued strength of Soviet operational art, which combined manoeuvre with attrition, mobile tank warfare with firepower. This can be seen in the conquest of Poland in the winter of 1944–5: the Red Army used large numbers of tanks, which were able to exploit opportunities prepared for by short and savage artillery attacks. The individual Soviet tank armies gained space to manoeuvre, and this prevented the Germans from consolidating new defensive positions. For forces that had broken through their opponents' lines, mobility enhanced the ability to prevent them from regrouping. The limit of the new advance was often linked to that of maintaining petrol supplies, as in the Red Army's advance through Poland in early 1945.

The pace of tank advance was also seen in Manchuria in August 1945 when Soviet armoured columns made rapid gains at the expense of Japanese forces. The Japanese underrated Soviet mobility.

The quality gap that favoured the Germans against the Anglo-Americans was closed by late 1944 and 1945, as new Allied tanks, particularly the heavily gunned American M-26 Pershing, appeared. The Americans also benefited in late 1944 from the introduction of high-velocity armour-piercing shells for their tank armament. This helped compensate for the earlier American emphasis on tanks that were fast and manoeuvrable.

WORLD WAR III

After World War II, tanks played a key role in preparations for conventional warfare between the Communist and Western blocs, with Soviet forces planning to unleash a large tank assault in Europe if World War III broke out, but their potential was not tested as conflict was avoided. Instead, the largest-scale use of tanks was in conventional warfare between second-rank powers. This was particularly the case in the Middle East. In the Six Day War, the Arab–Israeli conflict of 1967, the Israelis made good use of tanks. The war in Sinai was a large-scale tank conflict. Soviet T-54 and T-55 tanks used by the Egyptians were beaten by the American Patton and British Centurion tanks employed by the Israelis, who showed greater operational and tactical flexibility, not least in successfully searching out for vulnerable flanks, a task for which tanks were well suited, which thus overcame the strength of prepared Egyptian positions. Having broken into the Egyptian rear, the Israelis ably exploited the situation. The Egyptians retreated in chaos and Israel captured 320 tanks.

The conflict in Sinai also underlined the key role of field maintenance and repair in mobile warfare, the Israelis proving more effective than the Egyptians in every case. As always, overnight repair of equipment and its return to the battle line proved a key element in the war-making ability of a modern army. Tank engines are particularly prone to mechanical failure, while their tracks are also vulnerable.

In 1973, there was again large-scale tank conflict. Having captured the Israeli defences on the east bank of the Suez Canal on 6 October, the Egyptians repelled a series of Israeli counter-attacks, inflicting serious damage on Israeli armour. The Israelis suffered from a belief based on the experience of 1967, which exaggerated the effectiveness of tank attack and failed to provide adequate combined-arms capability, especially sufficient artillery support and mobile infantry, a lesson learned from tank conflict in World War II.

In response to Syrian pressure for help, however, the Egyptians changed their strategy and moved their armoured reserve forward, attacking on 14 October. This was a mistake, because, no longer taken by surprise, the Israelis were strong in defence. Their tactics were better as were their tanks: the Israelis' American-made M-48 and M-60 tanks had double the rate and range of fire of the Soviet T-55 and T-62 tanks. The Egyptians lost about 260 tanks that day. As in 1967, the failure of Arab armies in 1973 demonstrated the susceptibility of forces with a lower rate of activity, the problems arising from losing the initiative, and the need for a flexible defence. Israeli forces proved more effective in taking the initiative.

After World War II, tanks played a key role in preparations for conventional warfare between the Communist and Western blocs, with Soviet forces planning to unleash a large tank assault in Europe if World War III broke out, but their potential was not tested as conflict was avoided. Instead, the largest-scale use of tanks was in conventional warfare between second-rank powers.

In the 1973 war, command skills were particularly tested by the need to adapt to large-scale use of tanks. Attacks such as the Syrian advance with 800 tanks through the Israeli lines, and that by Egypt on 14 October, were clashes for which there was planning but scant experience. Egypt and Syria lost about 2,250 tanks in the war, the Israelis 840.

In the 1991 Gulf War, the Iraqis lost nearly 4,000 tanks to the American-led coalition forces. Aside from weapons' superiority, Allied fighting quality, unit cohesion, leadership and planning, and Iraqi deficiencies in each, all played a giant role in ensuring victory. In the 2003 Gulf War, the Iraqi T-55s and T-72s that were not destroyed by air attack could not prevail against the American Abrams tanks.

Atomic Bombs

'We knew the world would not be the same.
A few people laughed, a few people cried,
most people were silent. I remembered the
line from the Hindu scripture, the
Bhagavad-Gita. Vishnu is trying to persuade
the Prince that he should do his duty and to
impress him takes on his multi-armed form
and says, "Now, I am become Death, the
destroyer of worlds." I suppose we all
thought that one way or another.'

J. ROBERT OPPENHEIMER ON THE FIRST ATOMIC TEST

WORLD WAR II WAS BROUGHT TO A RAPID CLOSE IN 1945 when the dropping of two atomic bombs demonstrated that Japanese forces could not protect the homeland. At the Potsdam Conference, the Allies had issued the Potsdam Declaration, on the evening of 26 July, demanding unconditional surrender as well as the occupation of Japan, Japan's loss of its overseas possessions, and the establishment of democracy in the country. The alternative threatened was 'prompt and utter destruction', but, on 27 July, the Japanese government decided to ignore the Declaration.

Atom bombs were dropped on Hiroshima and Nagasaki on 6 and 9 August respectively. This transformed the situation, leading the Japanese, on 14 August, to agree to surrender unconditionally, although that also owed something to Soviet entry into the war on 8 August which removed any chance that the Soviets would act as mediators for a peace on more generous terms.

INTENSE RIVALRY

The creation of the atomic bomb was the culmination of an intense period of rivalry in invention and application which affected capability and combat. The atomic bomb was also indicative of the nature and scale of activity possible for an advanced industrial society. It was the product not only of the application of science, but also of the powerful industrial and technological capability of the US, and the willingness to spend about $2 billion in rapidly creating a large industry. The electromagnets needed for isotope separation were particularly expensive, and required 13,500 tons of silver.

The Germans and Japanese were both interested in developing an atomic bomb, but neither made comparable progress. The Uranverein, the German plan to acquire nuclear capability, was not adequately pursued, in part because the Germans thought it would take too long to develop. The German conviction that the war would, or could, be finished long before the bomb would be ready was encouraged by their military successes in 1939–41, but was an instance of over-confidence adversely affecting the development of new technologies. The Germans were also affected by hostility to what the Nazis termed 'Jewish physics', as well as the consequences of overestimating the amount of U235 required to manufacture a bomb.

In some respects, the use of atomic weapons suggested the obsolescence and indeed limitations of recent military practices. More people were killed in the American conventional bombing of Japan earlier in 1945 – the firebombing of Tokyo alone on 10 March, the first major low-level raid on the city, killing more than 83,000 people in one night – but that campaign required far more planes and raids: on 10 March, 334 B-29s were used, of which 14 were lost. Indeed, the use of atom bombs, like at a far more modest level that of jet aircraft by the Germans in the closing stages of the war in Europe, pointed the way towards a capability for war in which far fewer units were able to wield far more power. The one bomb dropped on Hiroshima, for example, destroyed 90 per cent of the city and – although it is impossible to be completely accurate – killed about 140,000. Some died from the effects of radiation in the weeks following the attack.

TOTAL DESTRUCTION

At the same time, the new use of atomic weapons in 1945 reflected not the limited capacity of pre-existing forms of warfare, but the extent to which they had created a military environment in which success was almost too costly. In short, a capacity for total destruction existed that would, it was hoped, now be short-circuited by modern warfare in the shape of the atomic bomb, the latter a logical consequence of the doctrine of strategic bombing of civilian targets as a means to an end.

The heavy Japanese and American losses on Iwo Jima, Okinawa and Luzon earlier in 1945 suggested that an Allied invasion of Japan, in the face of a suicidal determination to fight on, would be very damaging. The Japanese homeland army was poorly trained and equipped, and lacked mobility and air support, but, on the defensive, it would have the capacity to cause heavy casualties, particularly as it was unclear how to obtain the unconditional surrender that was an Allied war goal. General Douglas MacArthur remarked in April 1945 that his troops had not yet met the Japanese army properly, and that, when they did, they were going to take massive losses.

PRESIDENT TRUMAN

A rapid and complete victory seemed necessary in order to force Japan to accept terms that would neutralize its threat to its neighbours. It was also necessary to secure the surrender of the large Japanese forces in China and South-East Asia. The dropping of the atom bombs showed that the Japanese armed forces could not protect their people, and was therefore a major blow to militarists. A statement issued on behalf of President Harry Truman shortly after the first atomic bomb was dropped on Hiroshima declared,

> Hardly less marvelous has been the capacity of industry to design, and of labour to operate, the machines and methods to do things never done before, so that the brain child of many minds came forth in physical shape and performed as it was supposed to do... It was to spare the Japanese people from utter destruction that the ultimatum was issued at Potsdam. Their leaders promptly rejected that ultimatum. If they do not now accept our terms they may expect a rain of ruin from the air ... We are now prepared to obliterate more rapidly and completely every productive enterprise the Japanese have above ground in any city. We shall destroy their docks, their factories, and their communications. Let there be no mistake; we shall completely destroy Japan's power to make war.

Critics of American policy claim that the dropping of the bombs represented an early stage in the Cold War, with their use designed to ensure peace on American terms and both to show the Soviet Union the extent of American strength – in particular a vital counter to Soviet numbers on land – and to ensure that Japan could be defeated without the Soviets playing a major role.

This may have been a factor, but there seems little doubt that the prime use of the bombs was to avoid a costly invasion. Truman wrote on 9 August 'My object is to save as many American lives as possible, but I also have a human feeling for the women and children of Japan'. Had the war lasted until 1946, the destruction of the rail system by American bombing would have led to famine, as it would have been impossible to move food supplies. There were already systematic assaults on marshalling yards and raids. There would also have been more deliberate attacks on the cities. Aside from the raid on

Tokyo on 10 March 1955, there had been heavy raids on 13 and 19 April and 23 and 25 May. Similarly, there were heavy raids on Nagoya on 12 and 20 March and 14 and 16 May.

NUCLEAR EQUIVALENCE

America's nuclear monopoly, which appeared to offer a means to coerce the Soviet Union, lasted only until 1949. Then, thanks to successful spying on Western nuclear technology, the Soviet Union completed its development of an effective bomb that was very similar to the American one. This development had required a formidable effort, as the Soviet Union was devastated by the impact of World War II, and it was pursued because Joseph Stalin, the Soviet dictator, believed that only a position of nuclear equivalence would permit the Soviet Union to protect and advance its interests. However, such a policy was seriously harmful to the economy, as it led to the distortion of research and investment choices, and it was militarily questionable as resources were used that might otherwise have developed conventional capability.

Even when the US alone had the bomb, the value of the weapon was limited, as it was insufficiently flexible (in terms of military and political application or acceptance of its use) to meet challenges other than that of full-scale war. Thus, the US did not use the atom bomb (of which they then indeed had very few) to help their Nationalist Chinese allies in the Chinese Civil War, and their allies lost. Similarly, US possession of the bomb did not deter the Soviets from intimidating the West during the Berlin Crisis of 1948–9. Nevertheless, the availability of the bomb encouraged American reliance on a nuclear deterrent, which made it possible to hasten demobilization, leaving the USA more vulnerable when the Korean War (1950–3) broke out. In this conflict the American government decided in 1950 not to use atomic bombs. Instead, the war was fought with a strengthened conventional military force, although in 1953 the use of the atom bomb was threatened by the USA in order to secure an end to the conflict.

NUCLEAR STRATEGY

This encouraged the view that nuclear strategy had a major role to play in future confrontations, as indeed did the cost of fighting the Korean War, and the extent to which it had revealed deficiencies in the US military; although the war also caused a revival in the US army and led to its growing concern with readiness.

Meanwhile, as NATO countries were unable to match the build-up their military planners called for, there was a growing emphasis, especially from 1952, on the possibilities of nuclear weaponry both as a deterrent and, in the event of war, as a counterweight to Soviet conventional superiority.

The need to respond to Soviet conventional superiority on land and in the air encouraged an interest both in tactical nuclear weaponry and in the atom bomb as a weapon of first resort. The tactical nuclear weapons that were developed, such as bazookas firing atomic warheads with a range of one mile, were treated as a form of field artillery. The use of the atom bomb as a weapon of first resort was pushed by Dwight Eisenhower, NATO's first Supreme Allied Commander from 1950 until 1952, and US President from 1953 until 1961. Aware of NATO's vulnerability, he felt that strength must underpin diplomacy for it to be credible. In December 1955, the NATO Council authorized the employment of atomic weaponry against the Warsaw Pact, even if the latter did not use such weaponry. The potential of atomic weaponry was soon to be transformed by the development of ballistic missiles.

Ballistic Missiles

The means by which we live have outdistanced
the ends for which we live. Our scientific power
has outrun our spiritual power. We have
guided missiles and misguided men.'

IN 1957, THE SOVIET UNION LAUNCHED SPUTNIK I, the first satellite to go into orbit. The launch revealed a capability for intercontinental rockets that brought the entire world within striking range, and so made the US vulnerable to Soviet attack, both from first-strike and from counter-strike. In strategic terms, rockets threatened to give effect to the doctrine of airpower as a war-winning tool advanced in the 1920s and 30s, at the same time as they rendered obsolescent the nuclear capability of the bombers of the American Strategic Air Command, particularly the B-52s deployed in 1955.

The development of intercontinental missiles also altered the parameters of vulnerability, and ensured that space was more than ever seen in terms of straight lines between launching site and target. Nikita Khrushchev, the Soviet leader, declared in August 1961, 'We placed Gagarin and Titov in space, and we can replace them with bombs which can be diverted to any place on Earth'.

The threat to the US from Soviet attack was highlighted by the 1957 secret report from the American Gaither Committee. The strategic possibilities offered by nuclear-tipped long-range ballistic missiles made investment in expensive rocket technology seem an essential course of action, since they could go so much faster than aeroplanes and, unlike them, could not be shot down.

WERNER VON BRAUN

The USA had also been developing long-range ballistic missiles, using captured German V-2 scientists, particularly Wernher von Braun and many of his team from Peenemünde, and it fired its first intercontinental ballistic missile (ICBM) in 1958. The attempt to give force to the notion of massive nuclear retaliation entailed replacing vulnerable manned bombers with less vulnerable submarines equipped with ballistic missiles, and also with land rockets based in reinforced silos. The Americans fired their first intercontinental ballistic missile in 1958, and, in July 1960, off Cape Canaveral (subsequently Cape Kennedy), the USS *George Washington* was responsible for the first successful underwater firing of a Polaris missile. The following year, the Americans commissioned the USS *Ethan Allen*, the first true fleet missile submarine. Submarines could be based near the coast of target states, and were highly mobile and hard to detect.

They represented a major shift in the balance of power within the military forces, away from the US air force and towards the navy, which argued that its invulnerable submarines could launch carefully controlled strikes, permitting a more sophisticated management of deterrence and retaliation, an argument that was also to be made by the British navy.

BRITISH SUBMARINES

Other states followed the US. In 1962, in what became known as the Nassau Agreement, John F. Kennedy, the American president, and Harold Macmillan, the British prime minister, decided that the Americans would provide Polaris missiles for a class of four large nuclear-powered British submarines that were to be built, although American agreement was dependent on the British force being primarily allocated for NATO duties. In 1968, the first British Polaris test missile was fired from HMS *Resolution*, the British navy's first nuclear-powered ballistic missile submarine, which had been laid down in 1964. Polaris remained in service until 1995, being succeeded by Trident. The French commissioned their first ballistic missile submarine in 1969.

The risk of nuclear destructiveness made it important to prevent an escalation to full-scale total war, and encouraged interest in defining forms of warfare that could exist short of such escalation.

In the early 1960s, US concern about the nuclear balance increased. Kennedy had fought the 1960 presidential election in part on the platform that the Republican administration under Eisenhower had failed to maintain America's defences. Kennedy aimed for a strategic superiority over the Soviet Union and increased defence spending accordingly.

> Meanwhile, the destructive power of nuclear weapons increased when the atomic bomb was followed by the hydrogen bomb. The US first tested the bomb in 1952, destroying the Pacific island of Elugelab, and was followed by the Soviet Union in 1953, Britain in 1957, China in 1967, and France in 1968.

CUBAN MISSILE CRISIS

Concern about missiles rose to a peak during the Cuba crisis of 1962, when the Soviet Union deployed them in Cuba. These missiles had a range of 1,040 nautical miles, which made Washington a potential target. The Soviet intention was to protect Cuba from American attack, although it also shifted the balance of terror in the Soviets' favour. The USA imposed an air and naval quarantine to prevent the shipping of further Soviet supplies, prepared for an attack on Cuba, and threatened a full retaliatory nuclear strike. The Cuban leaders, Fidel Castro and Che Guevara, wanted a nuclear war, which they saw as a way to forward world socialism. However, the Soviet Union climbed down, withdrawing its missiles, while the US withdrew its Jupiter missiles (which carried nuclear warheads) from Turkey and agreed not to invade Cuba. Possibly the threat of nuclear war encouraged the US and the Soviets to caution, although both sides had come close to hostilities.

In the 1960s, both the US and the Soviet Union built up their missile forces. In 1965, Robert McNamara, the US Secretary of Defense, felt able to state that the US could rely on the threat of 'assured destruction' to deter a Soviet assault. Thanks in part to submarines, there would be enough missiles to provide an American counter-strike in the event of the Soviets launching a surprise first-strike and inflicting considerable damage.

The logic of deterrence, however, required matching any advance in the techniques of nuclear weaponry, and this was one of the most intense aspects of the Cold War. In 1970, the US deployed Minuteman III missiles equipped with multiple independently targeted re-entry vehicles (MIRVs), thus ensuring that the strike capacity of an individual rocket was greatly enhanced. This meant that any American counter-strike would be more effective. The US also cut the response time of their land-based intercontinental missiles by developing the Titan II, which had storable liquid propellants enabling in-silo launches, which reduced the launch time.

Meanwhile, the destructive power of nuclear weapons increased when the atomic bomb was followed by the hydrogen bomb. The US first tested the bomb in 1952, destroying the Pacific island of Elugelab, and was followed by the Soviet Union in 1953, Britain in 1957, China in 1967, and France in 1968. The hydrogen bomb employed a nuclear explosion to heat hydrogen isotopes sufficiently to fuse them into helium atoms, a transformation that released an enormous amount of destructive energy, far greater than that produced by a nuclear explosion. The precision of guidance systems was also increased.

MUTUALLY ASSURED DESTRUCTION

The American position in the 1970s was challenged by the Soviet response, part of the action-reaction cycle that was so important to the missile race. The Soviets made major advances in the development of land-based intercontinental missiles, producing a situation in which war was seen as likely to lead to MAD (mutually assured destruction), as both sides appeared to have a secure second-strike capability. Since the end of the Cold War, declassified Warsaw Pact documents have revealed that the Soviets then planned a large-scale use of nuclear and chemical weapons at the outset of any attack on Western Europe.

Concern about attack by ballistic missiles led to American interest in a 'Star Wars' programme. This was designed to enable the Americans to dominate space, using space-mounted weapons to destroy Soviet satellites and missiles. Subsequently, these ideas were to be applied in considering how best to challenge other states with long-range missiles, particularly North Korea.

The rise to power in the Soviet Union in 1985 of Mikhail Gorbachev, a leader committed to reform at home and good relations abroad, greatly defused tension. Gorbachev was willing to challenge the confrontational world-view outlined in KGB reports. For example, he was convinced that US policy on arms control was not motivated by a hidden agenda of weakening the Soviet Union, and this encouraged him to negotiate. In 1987, the Soviet government accepted the Intermediate Nuclear Forces Treaty, which, in ending land-based missiles with ranges between 500 and 5,000 kilometres (310 and 3,100 miles), forced heavier cuts on the Soviets, while also setting up a system of verification through on-site inspection. In 1991, START 1 led to a major fall in the number of American and Soviet strategic nuclear warheads.

Alongside developments in 'conventional' nuclear warheads, there was also investigation of the prospect for different warheads. One pursued by the Americans was the hafnium bomb which was seen as a way to produce a flood of high-energy gamma radiation. As the release of energy from the nuclei does not involve nuclear fission or fusion, such a bomb would not be defined as a nuclear weapon. So far, however, there have been major problems with such research. Concerned about the acquisition of weapons of mass destruction by 'rogue states', such as North Korea, the Americans seek a technological means to ensure security, a difficult goal at the best of times.

Cruise Missiles

'Cruise Missiles pose perhaps the gravest
delivery system proliferation threat. They
are inexpensive to build... they can penetrate
radar and infrared detection networks.
Finally since cruise missiles are unmanned,
they require no flight crew training,
expensive upkeep... or large air bases.'

FEDERATION OF AMERICAN SCIENTISTS

CRUISE MISSILES PROVIDED A KEY GAIN IN MILITARY CAPABILITY from the 1980s and were important to military planning in the last stage of the Cold War and to post-Cold War warfare. These missiles were valuable because they could deliver precise firepower without the risks and limitations associated with air power. In their planning for conflict with the Soviet Union in the 1980s, the US intended to respond to any attack by using cruise missiles to inflict heavy damage on advancing Soviet armour. These missiles can carry conventional warheads or use tactical nuclear weaponry. They can be fired in all weathers and can be launched from a variety of platforms.

The Soviet Union also developed such weaponry, but the arms race reflected the advantage the West enjoyed in electronic engineering, as well as higher growth rates and more flexible economic processes. However, for a long period, the Soviet belief in the apparently inevitably insoluble contradictions of Western capitalism ensured that they failed to appreciate the mounting crises their own economy, society and political systems were facing.

PRECISION BOMBING

Tactical nuclear warheads were not employed but in the Gulf War of 1991, cruise missiles and precision-guided bombs were employed by the US to provide precise bombardment. At that stage, cruise weaponry had a crucial advantage over air power because the US did not use many precision-guided munitions. Only 9,300 precision-guided munitions were dropped in that war and most of their aircraft were not equipped, or their pilots trained, for their use and, instead, employed unguided munitions, which made up 90 per cent of the aerial munitions used. This was despite the extensive and effective use of precision-guided munitions in the Linebacker I and II campaigns in Vietnam in 1972. The flexibility of cruise missiles was such that they could be launched from land, sea and air. Thus, the battleship USS *Wisconsin* was converted to ensure that it could launch missiles as well as fire guns in 1991.

Subsequently, the US fired 79 sea-launched cruise missiles at terrorist targets in Afghanistan and Sudan in 1998, an impressive but futile display of force which did not stop the terrorists. Indeed, Osama bin Laden was able to raise funds by selling missiles that did not detonate to the Chinese, who were interested in cutting-edge American military technology.

Cruise missiles were also used against Serbia in 1999 as part of a combined NATO air and missile assault designed to ensure that Serb forces withdrew from Kosovo. In 1998, the submarine HMS *Splendid* achieved Britain's first firing of a cruise missile, which had been bought from the US. The following year, *Splendid* fired these missiles at Serb targets in Kosovo as part of NATO operations there.

During the attack on the Taliban regime in Afghanistan in 2001, cruise missiles were fired from warships in the Arabian Sea. By then, air attack had improved as the availability of dual mode, laser and GPS guidance for bombs increased the range of precision available.

In the attack on Iraq in 2003, the precision of the cruise missiles and their attacks on Baghdad were presented as the cutting-edge of a shock and awe campaign that ushered in a new age of warfare. This was overstated and it still proved necessary to defeat Iraqi forces on the ground. The following year, there was speculation that a Chinese invasion of Taiwan would be countered by a Taiwanese cruise missile attack on the Three Gorges Dam in the Yangzi Valley, exploiting a key point of economic and environmental vulnerability.

Other states also developed cruise missile capacity. In 2004, Australia announced it would spend up to A$450 million on buying air-launched cruise missiles with a range of at least 250 kilometres (155 miles).

GLOBAL POSITIONING

Information in the form of precise positioning, was key to the effectiveness of the missiles. They made use of the precise prior mapping of target and traverse by satellites using a global positioning system, in order to follow predetermined courses to targets that were presented to weapons as grid references. The digital terrain models of the intended flight path facilitated precise long-distance firepower, while the TERCOM guidance system enabled course corrections to be made while in flight.

Such methods reflected the importance of complex automatic systems in advanced modern weapons. The force multiplier characteristics of weaponry had been greatly enhanced and become more varied. The mass production of the industrial age was replaced by technological superiority as a key factor in weaponry, not least because of the transformation of operational and tactical horizons by computers. This also encouraged a premium on skill which led to greater military concern about the quality of both troops and training. This encouraged military support for a professional volunteer force, rather than conscripts.

A good instance of this skill was the expertise required to direct UAVs (unmanned aerial vehicles) and RPVs (remotely piloted vehicles). These platforms are designed to take the advantage of using missiles a stage further by providing mobile platforms from which they can be fired or bombs dropped. Platforms do not require on-site crew and can be used without risk to the life or liberty of personnel. As a consequence, they can be low-flying, as the risk of losses of pilots to anti-aircraft fire has been removed. This is important given the extent to which the fate of captured pilots has, since the Vietnam War, become a major propaganda issue.

In addition, at least in theory, the logistical burden of air power is reduced. So also is the cost, as unmanned platforms are less expensive than manned counterparts, and there are big savings on pilot training. Unmanned platforms are also more compact and 'stealthy' (i.e. less easy to detect), while the acceleration and manoeuvrability of such platforms would no longer be limited by G-forces that would render a pilot unconscious.

> In the clash between Israel and Hizbollah in Lebanon in 2006, both sides used drones, with the Israelis making particularly marked use of them as an aspect of their aerial dominance and attack capacity. The difficulties encountered by Israel however indicated the contrast between force projection and military output, which missiles have greatly enhanced, and, on the other hand, being able to predict a successful resolution of the crisis. This is a universal problem with all weaponry.

UNARMED DRONES

In 1999, unarmed American drones were used extensively for surveillance over Kosovo in order to send information on bomb damage and refugee columns; and, in Afghanistan in 2001 and Iraq in 2003, armed American drones were used as firing platforms. The 26-foot American Predator with its operating radius of 500 miles, flight duration of up to 40 hours, cruising speed of 130 kph (80 mph), and normal operating altitude of 4,600 metres (aprox 15,000 feet), is designed to destroy air-defence batteries and command centres. It can be used in areas contaminated by chemical or germ warfare. In the clash between Israel and Hizbollah in Lebanon in 2006, both sides used drones, with the Israelis making particularly marked use of them as an aspect of their aerial dominance and attack capacity. The difficulties encountered by Israel, however, indicated the contrast between force projection and military output, which missiles have greatly enhanced, and, on the other hand, being able to predict a successful resolution of the crisis. This is a universal problem with all weaponry.

The background to all of this is major proliferation in the advanced weaponry held by a number of states in the late 1990s and early 2000s. This was particularly acute in South Asia. First India and then Pakistan tested nuclear weapons in 1998. That year, Pakistan also test-fired its new Ghauri intermediate-range missile, while India fired its new long-range Agni 2 missile the following year: its range is 2,000–3,000 kilometres (1,250–1,850 miles), extending to Tehran, and covering most of China and South-East Asia. In March 2003, both states test-fired short-range surface-to-surface missiles that could have been used to carry nuclear warheads. Pakistan, in turn, sold weapons technology to other states, including North Korea, Iraq, Iran, Libya and, probably, Egypt and Syria. Saudi Arabia probably funded the Pakistani nuclear-weapons programme and in the late 1980s purchased long-range Chinese missiles. In 2003, Iran conducted what it termed the final test of the Shahabz missile, first tested in 1998. With a range of 1,300 kilometres (812 miles), it is able to reach both Israel and US forces located in the region.

These weapon programmes were designed to provide regimes with the ability to counter the military superiority or plans of other states. Thus, North Korea saw atomic weaponry as a counter to American power, while Syria sought to develop chemical and biological weapons in response to Israeli conventional superiority. Japan, in turn, felt threatened by North Korea's rocketry, leading, in response, to Japanese interest in anti-missile defences and in satellite surveillance, while the Israelis built up a substantial stockpile of nuclear bombs in response to the chemical weapons of its Arab neighbours. In 2003, Libya abandoned its nuclear programme, but Iran proved unwilling to follow suit even after its violations of nuclear safeguards were exposed. In 2006, this caused a serious crisis in relations between Iran and the West, with Iran unwilling to back down in the face of international pressure. The advantages the Americans enjoyed with cruise missiles and other technology did not lead to the policy outcomes they had sought.

Index

Captions

FOR MIKE MOSBACHER

Quercus Publishing plc
21 Bloomsbury Square
London
WC1A 2NS

First published 2007

ISBNs
Cloth case edition   1 84724 228 6
                            978 1 84724 228 0
Printed case edition   1 84724 012 7
                            978 1 84724 012 5
Paperback   1 84724 220 0
                            978 1 84724 220 4

Printed and bound in Singapore

10 9 8 7 6 5 4 3 2 1

Picture credits

The publishers would like to thank the following for permission to reproduce illustrations and photographs:
5 R Sheridan/Ancient Art & Architecture Collection; 9 Visual Arts Library,(London)/Alamy; 13 Archaeological Museum Palermo/Dagli Orti (A)/Art Archive; 17 Egyptian Museum Cairo/Dagli Orti/Art Archive; 21 British Library/AKG – Images; 25 Araldo De Luca/Corbis UK Ltd; 29 Museo della Civilta Romana Rome/Dagli Orti/Art Archive; 33 Erich Lessing/AKG – Images; 37 Hazem Palace Damascus/Dagli Orti/Art Archive; 41 Visual Arts Library (London)/Alamy; 45 Mary Evans Picture Library; 49 Cott Aug A V f.51v Building of Marseilles/British Library, London, UK/Bridgeman Art Library; 53 Bettmann/Corbis UK Ltd; 57 AKG – Images; 61 Mary Evans Picture Library/Alamy; 65 Historical Picture Archive/Corbis UK Ltd; 69 Brian Seed/Alamy; 73 Mary Evans Picture Library/Alamy; 77 Erich Lessing/AKG – Images; 81 Mary Evans Picture Library; 85 Musée de la Marine Paris/Dagli Orti/Art Archive; 89 Historical Picture Archive/Corbis UK Ltd; 93 Francis G. Mayer/Corbis UK Ltd; 97 MPI/Hulton Archive/Getty Images; 101 Bettmann/Corbis UK Ltd; 105 Alinari/TopFoto; 109 Visual Arts Library (London)/Alamy; 113 Hulton-Deutsch Collection/Corbis UK Ltd; 117 Popperfoto/Alamy; 121 Mary Evans Picture Library/Alamy; 125 Popperfoto/Alamy; 129 Bettmann/Corbis UK Ltd; 133 TopFoto; 137 Bettmann/Corbis UK Ltd; 141 Mary Evans Picture Library; 145 Mary Evans Picture Library; 149 Hulton-Deutsch Collection/Corbis UK Ltd; 153 Ullstein bild/AKG – Images; 157 Popperfoto/Alamy; 161 G. Ttert/DPA/Corbis UK Ltd; 165 Hulton-Deutsch Collection/Corbis UK Ltd; 169 Bettmann/Corbis UK Ltd; 173 Bettmann/Corbis UK Ltd; 177 AKG – Images; 181 Corbis UK Ltd; 185 Bettmann/Corbis UK Ltd; 189 AKG – Images; 193 Bettmann/Corbis UK Ltd; 197 Ullstein bild/AKG – Images; 201 Kenneth Moll/US Navy/EPA/Corbis UK Ltd.

Edited and typeset by Windrush Publishing Services,
12 Adlestrop, Moreton in Marsh,
Gloucestershire GL56 0YN

Picture research by Zooid Pictures Limited
Index by Ingrid Lock